T0205403

Modeling and Optimization in Green Logistics

Houda Derbel • Bassem Jarboui • Patrick Siarry
Editors

Modeling and Optimization in Green Logistics

 Springer

Editors
Houda Derbel
Department of Quantitative Methods
and CS
University of Carthage
Nabeul, Tunisia

Bassem Jarboui
Department of Business
Higher Colleges of Technology
Abu Dhabi, United Arab Emirates

Patrick Siarry
Laboratoire LiSSi (EA 3956)
Université Paris-Est Créteil
Créteil, France

ISBN 978-3-030-45310-7   ISBN 978-3-030-45308-4   (eBook)
https://doi.org/10.1007/978-3-030-45308-4

This Springer imprint is published by the registered company Springer Nature Switzerland AG.
The registered company address is: Gewerbestrasse 11, 6330 Cham, Switzerland

Preface

The importance of green logistics to society and to competitive advantage continues to grow. Hence, industry and decision-makers have growing interest in improving the environmental aspects of transportation, one of the most important components of logistics and economic activity in the world. Despite the recent work in this field, the operational research side has remained weak. To that effect, the book proposes new optimization models suitable for green-vehicle routing problems, providing critical new perspectives of smart cities. These models are developed with respect to the economy and the logistics of supply chains.

Many variants of routing problems and different solution approaches including heuristics and metaheuristics are motivated by the impact of the solution on the environment and on human health. Solving related problems using optimization techniques and developing new models of network to access this energetic target have become a priority. This book presents recent works that analyse general issues of green logistics and smart cities. It includes original studies and findings in green logistics using promising techniques of operation research to achieve new operating models with ecological, economic and social objectives.

The book comprises green logistics systems including optimization models in the scope of the term. To that purpose, a literature review on routing problems and transportation is included to understand the different characteristics of green-vehicle routing problems. Then, new optimization approaches dealing with interesting aspects for sustainable supply chains with a view to reducing negative impacts on transportation and ensuring sustainability are exposed.

The main concern of this book is to present green transportation from both optimization and operational points of view. In addition to an exhaustive overview, the book investigates the use of operation research and optimization techniques for considering environmental issues within transportation logistics problems and solving different variants of green routing problems. It covers different ranges of problems and applications and emphasizes the green optimization logistics side.

To the best of our knowledge, there are few books dealing with the green logistics management and there is no book that focuses on the operational research perspective. Therefore, this book gives a new vision of what green logistics requires, and it focuses on the interrelationship between optimization and green transportation.

Organization of the Book

This book is organized into eight chapters. A brief description of each chapter is given below.

The chapter "The Green-Vehicle Routing Problem: A Survey" by G. Macrina, L.Di.P. Pugliese and F. Guerriero presents a survey of the main contributions related to the green-vehicle routing problem (G-VRP) and provides an up-to-date classification of the G-VRP variants presented in the literature while discussing the proposed solution approaches during the period from 2011 to 2019. The survey is based on two main classes, namely the G-VRP with conventional vehicles and the G-VRP with alternative fuel vehicles. An overview of variants and solution approaches addressed in each class is given.

The chapter "An Integrated Location-Inventory Routing Problem for ATMs in Banking Industry: A Green Approach" by N. Nazari-Ganjeei and S.M.J. Mirzapour-Al-E-Hashem presents a new mathematical model to integrate the location and vehicle routing problems for automated teller machines (ATMs) in the banking industry. The problem consists of choosing the optimal location for these facilities in addition to an optimal weekly routing for their cash replenishment. Knowing the location of the new facilities, the model concurrently attempts to provide an optimal cash replenishment policy by embedding a green-vehicle routing problem and taking into account a central warehouse and several types of banknote, in order to minimize total costs, including the transportation, deposition and shortage costs. The authors show that the model can be effectively used by banks to increase the performance of the ATM network using a real case study.

The chapter "Modelling a Future Routing Concept for Urban Air Mobility" by M. Hildemann and C. Delgado answers how to manage the air space in Manhattan properly by designing a model that satisfies citizens and relieves the urban transport system and also the environment. The authors then improve the model to be adaptable to different types and locations of urban areas and to temporal changes within each urban area. These temporal conditions are influenced by changing safety requirements, changing customer demand or changing noise level. This study demonstrates with a case study in New York City how a flight network for Urban Air Mobility can be designed to meet safety requirements and to increase the probability of acceptance by citizens for a new green mode of transportation.

The chapter "Putting the SC in SCORE: Solar Car Optimized Route Estimation and Smart Cities" by M. Hasicic and H. Siljak provides a solar car optimized route estimation (SCORE) based on historic and current solar radiance measurements.

This chapter deals with the relationship between SCORE and smart power generation and distribution systems (smart grid), novel transportation paradigms and communication advancements. The authors argue that SCORE will complement other smart transport solutions in a natural fashion, without disruptions and deadlocks in optimization.

The chapter "Evaluation and Prioritisation of Green Logistics and Transportation Practices Used in the Freight Transport Industry" by A. Kumar and R. Anbanandam develops a multi-criteria decision-making (MCDM) framework for evaluating green freight transport practices (GFTPs) used by the freight transport companies in an emerging economy (such as India). The result of the proposed framework provides a rank of important green logistics practices as well as low-performing practices. The fuzzy best–worst method (FBWM)-based analysis reported that competitive pressure from other freight transport companies is an important factor for adopting green transport practices. The results involve managerial implications for logistics managers.

The chapter "A Novel Hybrid Multi-objective Optimization Approach for Sustainable Delivery Systems with a Case Study in Izmir" by H.G. Resat presents a novel two-stage solution method designed for sustainable last-mile delivery systems in urban areas. A proposed hybrid solution methodology includes a multi-criteria decision-making system to select the most efficient logistics providers. Different performance indicators and a mixed integer linear programming model are considered. This model is being developed for drone based routing problem with time windows for last-mile delivery systems. The proposed solution methodology is applied in an illustrative case by using the real-life data of one metropolitan area in Turkey.

The chapter "When Green Technology Meets Optimization Modeling: The Case of Routing Drones in Logistics, Agriculture, and Healthcare" by M. Ndiaye, S. Salhi and B. Madani addresses the importance of drone technology as a green solution in the areas of logistics, agriculture and healthcare. A review of the recent applications of drone technology is presented to derive the resulted operational challenges, and variants of vehicle routing problems are presented as potential solutions for addressing the operational challenges of drone technology.

The chapter "Routing Electric Vehicles with Remote Servicing" by R.K. Arakaki, L.P. Maziero, M.D. Andrade, V.M.F. Hama and F.L. Usberti introduces the electric capacitated covering tour problem (ECCTP), a variant of the vehicle routing problem that allows customers' demands to be serviced remotely by electric vehicles with limited autonomy that echarge at alternative fuel stations (AFSs). The authors propose a mixed integer linear programming (MILP) mathematical formulation and a biased random-key genetic algorithm (BRKGA) metaheuristic for the ECCTP. Computational experiments show the effectiveness of the proposed methods while providing useful information for the decision-making on transportation operated by electric vehicles.

Audience

This book is dedicated to researchers and master's and PhD students working on green routing problems, and infrastructure managers working in the environmental sector. It will help them with regard to the perspectives for optimizing green logistics.

Nabeul, Tunisia   Houda Derbel
Abu Dhabi, United Arab Emirates   Bassem Jarboui
Créteil, France   Patrick Siarry

Contents

Contributors

Ramesh Anbanandam Department of Management Studies, Indian Institute of Technology Roorkee, Roorkee, India

Matheus Diógenes Andrade Institute of Computing, University of Campinas, Campinas, Brazil

Rafael Kendy Arakaki Institute of Computing, University of Campinas, Campinas, Brazil

Carlos Delgado Universidade Nova de Lisboa, Lisbon, Portugal

Luigi Di Puglia Pugliese Department of Mechanical, Energy and Management Engineering, University of Calabria, Rende, Italy

Francesca Guerriero Department of Mechanical, Energy and Management Engineering, University of Calabria, Rende, Italy

Ceren Gultekin Ozyegin University, Industrial Engineering, Istanbul, Turkey

Vitor Mitsuo Fukushigue Hama Graduate School of Informatics, Nagoya University, Nagoya, Japan

Mehrija Hasicic International Burch University, Sarajevo, Bosnia and Herzegovina

Moritz Hildemann Institute of Geoinformatics, Westfälische Wilhelms-Universität, Münster, Germany

Aalok Kumar Indian Institute of Technology Roorkee, Roorkee, India

Giusy Macrina Department of Mechanical, Energy and Management Engineering, University of Calabria, Rende, Italy

Batool Madani Industrial Engineering Department, American University of Sharjah, Sharjah, United Arab Emirates

Lucas Porto Maziero Institute of Computing, University of Campinas, Campinas, Brazil

S. Mohammad J. Mirzapour-Al-E-Hashem Department of Industrial Engineering and Management Systems, Amirkabir University of Technology (Tehran Polytechnic), Tehran, Iran
Rennes School of Business, Rennes, France

Nader Nazari-Ganjeh Department of Industrial Engineering and Management Systems, Amirkabir University of Technology (Tehran Polytechnic), Tehran, Iran

Malick Ndiaye Industrial Engineering Department, American University of Sharjah, Sharjah, United Arab Emirates

Okan Orsan Ozener Ozyegin University, Industrial Engineering, Istanbul, Turkey

Hamdi Giray Resat Izmir University of Economics, Department of Industrial Engineering, Izmir, Turkey

Said Salhi Centre for Logistics & Heuristic Optimization (CLHO), Kent Business School, University of Kent, Canterbury, UK

Harun Siljak CONNECT Centre, Trinity College, The University of Dublin, Dublin, Ireland

Fábio Luiz Usberti Institute of Computing, University of Campinas, Campinas, Brazil

Acronyms

2E-LRP	Two echelon-location routing problem
AFS	Alternative fuel station
AFV	Alternative fuel vehicle
ALNS	Adaptive large neighborhood search
APRS	Automatic packet reporting system
ATM	Automated teller machine
AUGMECON	The augmented epsilon-constraint method
BRKGA	Biased random-key genetic algorithm
CAD	Computer aided design
CRVRP	Capacitated recharging vehicle routing problem
CSP	Covering salesman problem
DDP	Drone delivery problem
DEA	Data envelopment analysis
DVRP	Drones vehicles routing problems
ECCTP	Electric capacitated covering tour problem
ELECTRE	Elimination and choice translating reality
EV	Electric vehicle
E-VRP	Electric vehicle routing problem
E-VRPTW	Electric vehicle routing problem with time windows and recharge stations
FBWM	Fuzzy best–worst method
FGP	Fuzzy goal programming
GAMS	General algebraic modeling system
GFTP	Green freight transport practices
GHG	Greenhouse gas emissions
GIS	Geographic information system
GP	Goal programming
GRASP	Greedy randomized adaptive search procedure
GUI	Graphical user interface
G-VRP	Green-vehicle routing problem
HAC	Heuristic algorithm based on convolution

HEV-TSP	Hybrid electric vehicle traveling salesman problem
ILP	Integer linear programming
IRP	Inventory routing problem
LRP	Location routing problem
MCDM	Multi-criteria decision-making
MCLM	Maximal covering location model
m-CCTP	Multi-vehicle cumulative covering tour problem
m-CTP	Multi-vehicle covering tour problem
MILP	Mixed integer linear programming
MIP	Mixed integer programming
ODS	One-direction search method
PHEV	Plug-in hybrid electric vehicle
PRP	Pollution-routing problem
PROMETHEE	Preference ranking organization method for enrichment evaluation
PSO	Particle swarm optimization
PVRP	Periodic vehicle routing problem
RLT	Reformulation-linearization technique
RVRP	Recharging vehicle routing problem
SAC	Simulated annealing based on convolution
SBRP	School bus routing problem
SC	Smart city
SCORE	Solar car optimized route estimation
SPRP	Steiner pollution routing problem
T-DPRPTW	Time-dependent pollution routing problem with time windows
T-DVRP	Time-dependent vehicle routing problem
TOPSIS	Technique for order preference by similarity to ideal solution
TSP-MD	Travelling salesman problem with a moving depot
UAV	Unmanned aerial vehicle
UPS	United parcel service
UTADIS	Utilities additives discriminants
VMND	Variable MIP neighborhood descent
VNS	Variable neighborhood search
VRPD	Vehicle routing problem with drones
VRPOD	Vehicle routing problem with occasional drivers
WCO	Waste cooking oil

Chapter 1
The Green-Vehicle Routing Problem: A Survey

Giusy Macrina, Luigi Di Puglia Pugliese, and Francesca Guerriero

Abstract In recent years, we have witnessed a dramatic rise of pollution levels in many areas of the world. Even if several green initiatives have been made in order to preserve and restore the environment, several nations do not respect their air quality standards. Due to the major impact that traffic has on air quality, the need to provide sustainable transportation plans is the main objective of many countries. We present a survey of the main contributions related to the green-vehicle routing problem (G-VRP). The G-VRP is a variant of the well-known vehicle routing problem, which takes into account the environmental sustainability in freight transportation. The main objective is to provide an up-to-date classification of the G-VRP variants presented in literature and discuss the proposed solution approaches.

1.1 Introduction

Green logistics aims at bringing the environmental perspective in traditional logistics (see [76]). In recent years, governments and business organizations have triggered several green initiatives, as a result, interest in green logistics has increased as well as the society's environmental awareness. In this perspective, due to the great impact that transportation logistics has on the environment, reducing negative externalities in transportation logistics is one of the main goals for many countries. The vehicle routing problem (VRP), introduced by Dantzig and Ramser [18], aims at finding the optimal delivery/collection routes for a fleet of vehicles from a depot to a set of customers. The VRP often includes some traditional constraints, such

G. Macrina (✉) · F. Guerriero
Department of Mechanical, Energy and Management Engineering, University of Calabria, Rende, CS, Italy
e-mail: giusy.macrina@unical.it; francesca.guerriero@unical.it

L. Di Puglia Pugliese
Istituto di Calcolo e Reti ad Alte Prestazioni, Consiglio Nazionale delle Ricerche, Rende, CS, Italy
e-mail: luigi.dipugliapugliese@icar.cnr.it

© Springer Nature Switzerland AG 2020
H. Derbel et al. (eds.), *Modeling and Optimization in Green Logistics*,
https://doi.org/10.1007/978-3-030-45308-4_1

1

as capacity, route length, time windows, precedence relations between customers, etc. (see Laporte [54]). Since the VRP is a central problem in freight transportation, it was widely studied during the years (see Laporte [53], Laporte [54] and Kumar and Paneerselvam [52]). Recently, several authors started to study the VRP under a green perspective, by considering environmental effects of routing strategies, use of alternative fuel vehicles, energy minimization, etc.. We call this VRP variant green-vehicle routing problem (G-VRP). Lin et al.[57] proposed a survey on G-VRP. They surveyed the main VRP variants, then they focused on green logistics contributions during 2006–2012. They classified the G-VRP in three main classes: green-VRP, pollution routing problem, and VRP in reverse logistics.

The aim of our work is to survey and classify the G-VRP variants introduced during 2011–2019, and describe the proposed solution approaches for these problems. The remainder of this paper is organized as follows. Section 1.2 presents the survey methodology. Section 1.3 gives an overview of the main scientific contributions regarding the G-VRP with conventional vehicles. In Sect. 1.4 we review the G-VRP with alternative fuel vehicles variants. Section 1.5 describes other innovative applications in VRP fields, concerning the use of alternative vehicles in the delivery process. Conclusions follow in Sect. 1.6

1.2 Survey Methodology

In order to review the literature on G-VRP, we used several academic databases, including Scopus, Google scholar, and Sciencedirect, accessed from the university library by using keywords such as green-vehicle routing, green logistics, and pollution routing problem. We searched papers in journals, books, technical reports, and conference proceedings. We also used the bibliographies of survey papers and papers on G-VRP. We reviewed 75 papers on G-VRP in the period fixed from 2011 to 2019. Figure 1.1 shows the distributions of the surveyed works during the years. In large part the contributions are scientific articles, published in operations research journal such as European Journal of Operational Research, Operations Research, Transportation Research (parts: B,C,E), Transportation Science; four are proceedings, only three are technical reports and one is an unpublished article. We propose a classification scheme based on the different variants of G-VRP presented in scientific literature. In particular, we identified two main classes: (1) the G-VRP with conventional vehicles and (2) the G-VRP with alternative fuel vehicles. We also identified four sub-categories for the G-VRP with conventional vehicles and six sub-categories for the G-VRP with alternative fuel vehicles (see Fig. 1.2). We describe the variants and the proposed approaches for these problems. Since the VRP is classified as NP-hard problem, solving the G-VRP with exact optimization methods may be very difficult. Only eight out of 75 works proposed an exact method to solve the proposed G-VRP variant. Large part of the authors proposes local search based metaheuristics for their problems, such as adaptive large neighborhood search, variable neighborhood search, and tabu search, which look highly performing for this class of problems.

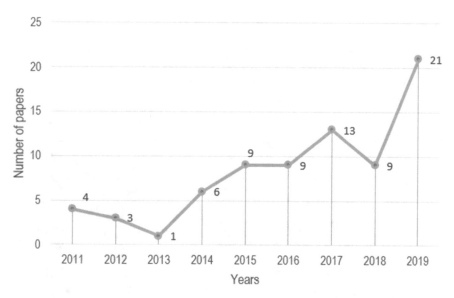

Fig. 1.1 Number of contributions during the years

1.3 Literature Review on the G-VRPs with Conventional Vehicles

Externalities in freight transportation are various. The CO_2 emissions problem is one of the most known and significant, due to the negative impact on the environment and human health. Several authors explicitly consider the CO_2 emissions in the objective function, and focus on minimization of routing cost and polluting emissions.

Bektaş and Laporte [9] introduced and modeled the Pollution Routing Problem (PRP). The PRP explicitly considers the polluting emissions in the objective function for the first time. The authors modeled an energy consumption of conventional vehicles and presented a non-linear mixed integer mathematical problem for the PRP. Starting from this paper, several authors proposed different modeling and algorithmic extensions. Figure 1.3 shows the trend of publication from 2011 to 2019.

Demir et al. [20] showed the difficulty of solving the PRP, by using the model presented in Bektaş and Laporte [9]. The authors proposed an extension of the PRP and an adaptive large neighborhood search (ALNS) heuristic capable of solving instances with up to 200 nodes. Kramer et al. [51] developed a new hybrid iterated local search (ILS) that integrates a set partitioning procedure and a speed optimization algorithm for the PRP addressed in Bektaş and Laporte [9]. Their approach highly outperforms the previous available algorithms. The authors considered the same speed on each arc and assumed that the departure time is fixed. However, there exists an optimal speed yielding a minimum fuel consumption (see Demir et al. [20]), which leads to a minimization of CO_2 emissions; thus,

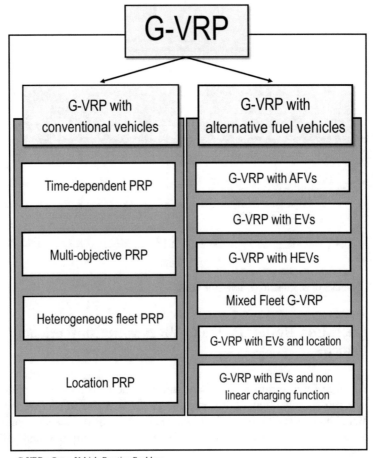

G-VRP = Green Vehicle Routing Problem
PRP = Pollution Routing Problem
AFVs = Alternative Fuel Vehicles
EVs = Electric Vehicles
HEVs = Hybrid Electric Vehicles

Fig. 1.2 Classification of contributions

the main goal for the PRP is the optimization of the speed for each route. For this reason, Kramer et al. [50] extended the previous work (i.e., Kramer et al. [51]), by introducing variable departure times, and embedding speed and departure time in the optimization algorithm proposed in Kramer et al. [51]. De Oliveira da Costa et al. [19] proposed a genetic algorithm and tested it on both benchmark and real-world instances. Majidi et al. [64] presented a non-linear mixed integer programming model for the PRP with simultaneous pickup and delivery. Their goal was to minimize fuel consumption and emissions by scheduling and routing customers. They presented an ALNS heuristic.

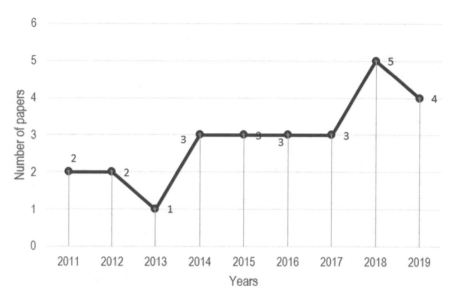

Fig. 1.3 Number of PRP contributions during the years

1.3.1 Time-Dependent PRP

Figliozzi [31] introduced a time-dependent VRP with time windows (T-DVRPTW). The author calculated the amount of fuel spent and studied the impacts of travel speed, congestion, and land use on CO_2 emissions. Jabali et al. [42] presented a tabu search for solving the T-DVRP, considering maximum achievable vehicle speed as a part of the optimization. They considered a two-stage planning horizon: (1) free flow traffic and (2) congestion. They modeled and minimized the emissions per kilometer as a function of speed, and showed the relation between the reduction of emissions and costs. Franceschetti et al. [32] extended the PRP by explicitly considering the effect of traffic congestion. Hence, they studied the T-DPRPTW. They proposed an integer linear programming formulation and partitioned the planning horizon into two phases, as in Jabali et al. [42]. Huang et al. [98] extended the work of Franceschetti et al. [32] considering path selection in the road networks as an integrated decision in the T-DVRP. This problem is denoted as T-DVRP with path flexibility. Introducing path flexibility means that any arc between two customers has multiple corresponding paths in the road network. The authors modeled the T-DVRP with path flexibility under stochastic traffic conditions and precomputed a set of candidate paths between every pairs of customers. However, they did not guarantee that all the eligible paths are considered. Another variant of the problem addressed by Franceschetti et al. [32] is that considered by Ehmke et al. [27]. The authors extended the T-DVRP considering more than two periods, not allowing vehicles to wait, and taking into account for the variability in travel times. In addition, they assumed that vehicles must travel at the speed of traffic, which can vary; hence, time-dependent speeds are not deterministic. One of the major limitations of most of the existing research works is that they consider a-priori determined single-road path for traveling between each pair of customers. However, the expected emissions-minimized path between any two customers can vary, thus it is not guaranteed that it is possible to identify all the paths a-priori as in most VRPs. To address this issue, Ehmke et al. [27] proposed an approach to find emissions-minimized routes, identifying a condition under which a time-dependent path between two nodes is load invariant. This result allowed them to reduce computational challenge of finding time- and load-dependent paths between some customers. However, for the other node pairs where this condition is not satisfied, they needed to compute the time- and load-dependent path within the routing algorithm, which is a tabu search. They solved instance with up to 30 customers. Raeesi et al. [78] introduced the Steiner PRP (SPRP) that is a multi-objective, time- and load-dependent PRP with a heterogeneous fixed fleet, time windows, flexible departure times, and multi-trips on congested urban road networks. The SPRP minimizes three objective functions: vehicle costs, amount of fuel, duration of the routes. Contrary to the previous works, the proposed SPRP works directly with an instantaneous emissions model formula, hence it incorporates second-by-second speed variations and acceleration/deceleration rates. Starting from the work of Ehmke et al. [27], the authors developed a path elimination procedure (PEP) to discard all redundant paths between two customers in a pre-processing phase, and proposed a PEP-based mixed integer programming formulation of the SPRP to avoid the calculation of shortest paths and reduce the computational efforts. Then, they tested the performance of their model with CPLEX. Xiao and Konak [90] addressed the time-dependent vehicle routing and scheduling problem, with the purpose to minimize emissions and weighted tardiness. Tardiness objective is usually presented in scheduling problems for modeling the timeliness of a service, where it is not possible to serve all costumers on time because of capacity constraints. The authors modeled the problem of arcs routing and time scheduling, and presented an alternative formulation for modeling time dependency. They developed a genetic algorithm and a dynamic programming technique for solving the routing and scheduling problems, respectively. In particular, the genetic algorithm searches for the routing and vehicle selection decisions, and dynamic programming optimizes the scheduling decision of the selected vehicles and routes. They solved instances with up to 200 customers. Qian and Eglese [77] considered the problem of fuel optimization in VRP with time-varying speeds, which aims at minimizing the total fuel emissions. The speeds of the vehicles depend on time and must be determined, thus they are treated as decision variables. The authors proposed a column generation based tabu search algorithm to solve the problem, and tested it on real-life data from a London road network. Tajik et al. [86] introduced uncertainty in the T-DPRP with pickup and delivery. They defined a mixed integer linear program in which the main objective is to minimize the travel distance, the number of vehicles, and the polluting emissions. Then, they introduced a robust extension, considering the vehicle speed as an uncertain parameter.

1.3.2 Multi-Objective PRP

Demir et al. [21] proposed and solved the bi-objective PRP, in which they minimized two conflicting factors: fuel consumption and driver time. The problem was solved via a bi-objective ALNS algorithm combined with a speed optimization procedure, which consists of computing the optimal speed on each arc of the route in order to minimize the objective function. The computational results showed that it is not needed to increase the driving time significantly in order to reduce fuel consumption and CO_2 emissions. Starting from the work of Demir et al. [21], Suzuki [85] presented a "practical" model of PRP focusing on the users' point of view. In particular, the authors selected only a subset of the factors considered by Demir et al. [21] to calculate the fuel consumption of vehicles. They proposed a simulated annealing and a tabu search to solve their problem. As in Demir e al. [21], Costa et al. [17] considered a bi-objective PRP. The authors proposed a multi-objective approach based on the two-phase Pareto local search heuristic to solve the problem. They showed that their proposed algorithms outperform the method used by Demir et al. [21] to solve the same problem; in fact, they obtained high-quality solutions in less CPU time. Rauniyar et al. [79] proposed a novel solution approach based on the well-known non-dominated sorting genetic algorithm-II to solve a multi-objective PRP derived by Demir et al. [21]. In order to improve the performance of their heuristic, the authors integrated an evolutionary algorithm, called multi-factorial optimization, which is able to optimize the distance of all possible routes simultaneously. Poonthalir and Nadarajan [75] modeled and solved a bi-objective PRP with varying speed constraints. The model minimizes both route cost and fuel consumption. The problem is solved using particle swarm optimization with greedy mutation operator and time varying acceleration coefficient. The authors carried out a comprehensive computational study using four different categories of speed intervals. Then, they estimated and analyzed the corresponding route cost and fuel consumption. Their results confirm that constant speed consumes more fuel than varying speed.

1.3.3 Heterogeneous Fleet PRP

Koç et al. [48] proposed a PRP variant with a heterogeneous fleet, named the fleet size and mix PRP. They proposed a hybrid evolutionary metaheuristic to solve it and showed that in an urban setting the benefits of using a heterogeneous fleet are superior to a homogeneous one. Kancharla and Ramadurai [44] focused on the effect of acceleration and deceleration, as well as load carried and speed, on fuel consumption for a heterogeneous fleet of conventional vehicles. They proposed a mixed integer programming formulation and solved small size instances with up to 33 customers. To solve a multi-objective heterogeneous pickup and delivery PRP, Bravo et al. [11] proposed an evolutionary algorithm.

They considered the following objective functions: CO_2 emissions measured in terms of fuel consumption, total traveling time of vehicles, and number of customers served. They used the presented model to solve small size instances; their results highlighted an inverse relationship between fuel consumption and travel time, due to the speed term present in both the objective functions. Yu et al. [94] extended the work of Koç et al. [48], hence modeled and solved the heterogeneous fleet green vehicle routing problem (HFGVRP) with time windows. They formulated the problem and developed several versions of a branch and price algorithm. They first considered a basic branch and price algorithm with the bidirectional labelling algorithm for the HFGVRP, and then developed an improved version of the algorithm to reduce labels and rapidly solve the problem.

1.3.4 Location PRP

Toro and Franco [87] presented a multi-objective location G-VRP. In particular, they modeled a bi-objective problem, considering the minimization of operational costs and the minimization of environmental impacts. Dukkanci et al. [26] modeled and solved the green location routing problem, which is a combination of location routing problem and PRP. This problem consists in locating depots where vehicle with limited capacity will be dispatched to serve a set of customers, setting the speed on each arc of the tour in order to satisfy their time windows, minimizing the overall costs and emissions. The authors proposed an integer programming based algorithm and an ILS heuristic to solve that problem.

Table 1.1 summarizes the main papers on PRP and its variants.

1.4 Literature Review on the G-VRPs with Alternative Fuel Vehicles

A different approach is to use alternative fuel vehicles (AFVs), especially electric vehicles (EVs), instead of the conventional ones. Governments have started to give incentives aimed at increasing the commercial use of EVs, (see Pelletier et al. [72]). Recently, Erdelić and Carić [28] proposed a survey on the electric vehicle routing problem (E-VRP). They surveyed variants and solution approaches for this class of problem. While EVs do not produce CO_2 emissions and are more silent than the conventional vehicles, they are constrained by the low autonomy of their battery, the limited number of public charging stations (CSs), and long charging times. In the last years, the number of publications on this topic considerably increased with the interest on green transportation. The scientific community started to study and

Table 1.1 Summary of the literature on the PRP and its variants

Reference	Algorithm	Mathematical model	Time windows	Time dependency	Pickup and delivery	Uncertain data	Heterogeneous fleet	Multi-objective	Location
Bektaş and Laporte [9]		•	•						
Figliozzi [31]	Heuristic	•	•	•					
Demir et al. [20]	Heuristic	•	•						
Jabali et al. [42]	Heuristic			•					
Franceschetti et al. [32]			•	•					
Demir et al. [21]	Heuristic	•	•						
Tajik et al. [86]		•	•		•	•		•	
Koç et al. [48]	Heuristic	•	•				•		
Kramer et al. [51]	Heuristic	•	•						
Kramer et al. [50]	Heuristic								
Qian and Eglese [77]	Heuristic	•	•	•					
Ehmke et al. [27]	Heuristic			•					
Xiao and Konak [90]	Heuristic	•	•	•			•	•	
Suzuki [85]	Heuristic	•							
Huang et al. [98]		•		•					
Toro and Franco [87]	Heuristic	•					•	•	•
Bravo et al. [11]	Heuristic	•	•				•	•	
Kancharla and Ramadurai [44]		•					•		

Reference	Method								
de Oliveira da Costa et al. [19]	Heuristic					●			
Majidi et al. [64]	Heuristic	●							
Costa et al. [17]	Heuristic	●					●		
Raumiyar et al. [79]	Heuristic						●		
Poonthalir and Nadarajan [75]	Heuritic	●					●		
Yu et al. [94]	Exact	●	●				●		
Raeesi et al. [78]	Heuristic	●	●	●			●	●	
Dukkanci et al. [26]	Heuristic	●	●				●		●

Table 1.2 Recent studies of G-VRP with AFVs during 2011–2019

G-VRP variants	Papers	Number
G-VRP with AFVs	[1, 4, 13, 14, 16, 29, 41, 47, 55, 67, 96]	11
G-VRP with EVs	[8, 22, 24, 30, 35, 39, 43, 45, 56, 59, 73, 83, 84, 97]	14
G-VRP with HEVs	[25, 38, 65, 99]	4
Mixed fleet VRP	[12, 36, 37, 40, 49, 62, 63, 80, 92, 93]	10
G-VRP with EVs and location	[58, 71, 81, 82, 91, 95]	6
G-VRP with EVs and non-linear Charging function	[33, 34, 46, 68]	4

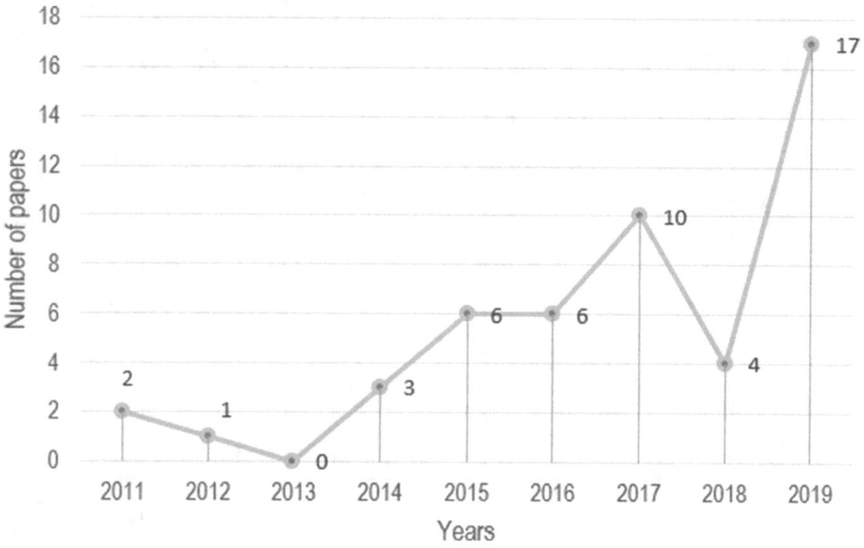

Fig. 1.4 Number of contributions in G-VRPs with AFVs published during the last years

propose several G-VRP with AFVs variants, in particular we have identified and classified five classes of this problem resumed in Table 1.2.

Figure 1.4 shows the trend in publications on G-VRP with AFVs and its variants during the last 8 years. Looking at Fig. 1.4, it is clear that a large number of scientific contributions were published during the years 2014–2019 with a growing trend, with the only exception of the year 2018. Table 1.3 summarizes the main papers addressing the G-VRP with AFVs and its variants.

1.4.1   Green-Vehicle Routing Problem with AFVs

Erdoğan and Miller-Hooks [29] for the first time used the name "G-VRP" to address this class of problem. They introduced the G-VRP in which the fleet is composed of AFVs. Fuel consumption is proportional to traveled distance, and the vehicle fuel tank can be charged at alternative fuel charging stations. The authors proposed

Table 1.3 Summary of the literature on the G-VRP with AFVs variants

	Reference	Algorithm	Mathematical model	Time windows	Fixed charging	Partial recharge	Location of CSs	Multiple technologies	Battery swap	Linear charging	Non-linear charging	Energy consumption rate	Energy consumption model	Pickup and delivery	Multi-depot	Mixed fleet
AFVs-VRP	Conrad and Figliozzi [16]	Heuristic	•	•						•		•				
	Erdoğan and Miller-Hooks [29]	Heuristics	•		•							•				
	Leggieri and Haouari [55]	Exact	•		•							•				
	Montoya et al. [67]	Heuristic			•							•				
	Koç and Karaoglan [47]	Exact			•							•				
	Affi et al. [1]	Heuristic										•				
	Hooshmand and MirHassani [41]	Heuristic	•													
	Adelmin and Bartolini [4]	Heuristic								•		•				
	Bruglieri et al. [14]	Heuristic and Exact	•							•		•				
	Bruglieri et al. [13]	Heuristic	•							•		•				
	Zhang et al. [96]	Heuristic										•			•	
	Schneider et al. [83]	Heuristic	•	•						•		•				
	Felipe et al. [30]	Heuristic	•			•		•		•		•				•
	Ding et al. [24]	Heuristic	•	•		•				•		•		•		
	Bruglieri et al. [59]	Heuristic	•	•						•		•				
G-VRP with EVs	Desaulniers et al. [22]	Exact	•	•		•				•		•				
	Lin et al. [56]	Heuristic	•										•			
	Hiermann et al. [39]	Heuristic and Exact	•	•						•		•		•		
	Keskin and Çatay [45]	Heuristic	•	•		•				•		•				
	Joo and Lim [43]	Heuristic	•									•				
	Zhang et al. [97]	Heuristic	•							•			•			
	Shao et al. [84]	Heuristic	•	•	•							•	•			
	Goeke [35]	Heuristic	•	•		•				•		•		•		
	Pelletier et al. [73]	Heuristic	•			•				•		•		•		

Category	Reference	Method
G-VRP with HEVs	Basso et al. [8]	
	Doppstadt et al. [25]	Heuristic
	Vincent et al. [38]	Heuristic
	Mancini [65]	Heuristic
	Zhen et al. [99]	Heuristic
	Gonçalves et al. [37]	
Mixed fleet G-VRP	Sassi et al. [80]	Heuristic
	Goeke and Schneider [36]	Heuristic
	Yavuz and capar [93]	Heuristic
	Yavuz [92]	Heuristic
	Macrina et al. [63]	Heuristic
	Macrina et al. [62]	Heuristic
	Breunig et al. [12]	Heuristic and Exact
	Hiermann et al. [40]	Heuristic
	Koyunku and Yavuz [49]	Heuristic
	Yang and Sun [91]	Heuristic
G-VRP with EVs and location	Li-ying and Yuan-bin [58]	Heuristic
	Schiffer and Walther [81]	Exact
	Schiffer and Walther [82]	Heuristic
	Paz et al. [71]	
	Zhang et al. [95]	Heuristic
G-VRP with EVs and non-linear charging function	Montoya et al. [68]	Heuristic
	Froger et al. [34]	Heuristic and exact
	Keskin et al. [46]	Heuristic
	Froger et al. [33]	Heuristic

two constructive heuristics for the problem, with the goal of minimizing the traveled distance. Conrad and Figliozzi [16] presented the recharging VRP with time windows, where the vehicles have a limited driving range and have to be charged to continue their route. Recharging at some customer locations is allowed. The authors proposed an iterative construction and improvement heuristic to solve the problem. Montoya et al. [67] proposed a multi-space sampling heuristic to solve the G-VRP with AFVs proposed in Erdoğan and Miller-Hooks [29]. The procedure includes three main components: three randomized traveling salesman problem heuristics, a tour partitioning procedure, and a set partitioning formulation. They compared their approach with those proposed by Erdoğan and Miller-Hooks [29] and that of Schneider et al. [83] which is an extension of the G-VRP with electric vehicles that will be introduced in Sect. 1.4.2, adapted by Montoya et al. [67] to their problem. They concluded that their heuristic is highly competitive, and also the simplest for the G-VRP. Also Koç and Karaoglan [47] and Affi et al. [1] studied the G-VRP with AFVs introduced by Erdoğan and Miller-Hooks [29]. In particular, Koç and Karaoglan [47] proposed a new formulation for the G-VRP with AFVs, introducing new decision variables to allow multiple visits to the CSs without augmenting the networks with additional dummy nodes. Then, they proposed a simulated annealing heuristic based on an exact solution approach. Their results showed that optimal solutions for 22 out of 40 test instances with 20 customers were obtained within reasonable computation time. Affi et al. [1] proposed a variable neighborhood search (VNS) algorithm, and solved instances with up to 500 customers. Zhang et al. [96] proposed an ant colony algorithm for the G-VRP with multiple depots and solved instances with up to 75 customers and six depots. Bruglieri et al. [14] proposed two formulations for the G-VRP with AFVs. In the first formulation, a vehicle can visit only one station between two customers, whereas in the second one two consecutive visits to the stations are permitted. They compared their formulations with those proposed by Erdoğan and Miller-Hooks [29] and Koç and Karaoglan [47]. Their computational results showed that their models outperform the already existing exact solution approach. Based on the work of Koç and Karaoglan [47], Leggieri and Haouari [55] proposed a new formulation for the G-VRP with AFVs. In order to assess the effectiveness of their approach, the authors solved the proposed model by using CPLEX and compared the results with those obtained by the branch-and-cut algorithm proposed in Koç and Karaoglan [47]. Bruglieri et al. [13] proposed a two-phase solution approach in which a route is represented as a composition of paths. They proposed an exact approach for small size instances and converted it into a heuristic approach for larger instances. They compared their heuristic with those of the literature [29, 47, 55] and showed its good performances. Adelmin and Bartolini [4] proposed a multi-start local search heuristic for solving the G-VRP with AFVs. They tested their heuristic on benchmark instances with up to 100 customers and compared the results with those obtained by the algorithms in the literature [29, 30, 67, 83]. They showed that their heuristic performs very well in terms of solution quality; however, even if it is not the fastest, the computational times are competitive. To consider the effect of the congestion on travel time and fuel consumption, Hooshmand and MirHassani [41] proposed a time-dependent G-

VRP with AFVs, in which the travel time between two nodes depends on distance traveled as well as the time of the day and the vehicle speed. After modeling the problem, they proposed a hybrid heuristic algorithm based on clustering techniques and simulated annealing framework.

1.4.2   Electric Vehicle Routing Problem

Interest in EVs has increased in recent years, due to government incentives and technology progress. Using EVs in good distribution is considered a serious alternative to the conventional vehicles, hence, several authors started to study the VRP with fleets composed of EVs and its variants. Schneider et al. [83] extended the work of Erdoğan and Miller-Hooks [29] by introducing the E-VRP with time windows and recharge stations (E-VRPTW). EVs can be charged at any of the available CSs, and charging time is related to the battery charge level when the vehicle arrives at the CS. They proposed a hybrid metaheuristic that integrates VNS with tabu search. Also Felipe et al. [30] extended the model presented in Erdoğan and Miller-Hooks [29]. They allowed partial recharges and considered multiple charging technologies at CSs. The authors developed a nearest neighbor construction heuristic, as well as a simulated annealing algorithm. Zhang et al. [97] proposed an ant colony algorithm based metaheuristic to solve the E-VRP. Ding et al. [24] and Goeke [35] addressed the E-VRPTW with partial recharges and the pickup and delivery policy. Ding et al. [24] extended the E-VRPTW model of Schneider et al. [83], considering also limited chargers at a charging station and flexibility of partial charging for EVs, and proposed a hybrid heuristic which incorporates VNS with a tabu search. Goeke [35] presented a compact formulation based on recharging paths for the problem and developed a granular tabu search heuristic to solve it. Bruglieri et al. [59] studied a variant of E-VRPTW in which the battery charging level is a decision variable. They solved the problem with a VNS branching. Desaulniers et al. [22] developed two branch-prince-and-cut algorithms for the E-VRPTW and extended the problem by considering four charging strategies: a single charge or multiple charges per route and fully recharge only, multiple recharges per route and batteries are fully charged, at most a single recharge per route and partial recharges, and multiple partial recharges. Lin et al. [56] extended the E-VRP by considering a heterogeneous fleet of EVs and the vehicle load effect on battery consumption proposed by Goeke and Schneider [36]. They solved the model with CPLEX and compared alternative routing strategies using a case study in Austin, Texas. Hiermann et al. [39] introduced the electric fleet size and mix vehicle routing problem with time windows and charging stations. They considered a heterogeneous fleet of EVs in which each vehicle is characterised by its fixed cost, battery and load capacity, energy consumption, and charging rate. Each vehicle can be fully charged at a CS. They proposed a hybrid metaheuristic which combines ALNS to a cyclic neighborhood search and labeling procedures. Keskin and Çatay [45] formulated the E-VRPTW with partial recharges and solved it with an ALNS procedure. The authors showed the benefits of partial recharges on total costs, compared to a full recharge strategy. Shao et al. [84] considered the time-dependent context, introducing the E-VRPTW with charging time and variable travel time. They fixed the charging time to 30 min and batteries are always full charged. The authors proposed a dynamic Dijkstra algorithm to find the shortest path between two nodes in the routes, and a genetic algorithm to obtain the routes. Joo and Lim [43] proposed an ant colony optimized routing strategy to solve the E-VRP, considering recuperation energy and speed of vehicles. In their study, the authors did not consider the possibility to charge the battery. Energy consumption is one of the main issues facing of the E-VRP. Indeed, many factors such as: weather, road condition, or driver behaviour can affect the consumption. Hence, Pelletier et al. [73] introduced the E-VRP with energy consumption uncertainty. They formulated the problem as a robust mixed integer linear program and solved small size instances. For larger instances they proposed a two-phase heuristic method based on large neighborhood search. Basso et al. [8] introduced a two-stage E-VRP that incorporates energy consumption estimation by considering topography and speed profiles. The authors presented a two-step approach with energy estimation approximated driving cycle considering speed, acceleration and braking, to estimate energy demand into the routing problem.

1.4.3 Hybrid Vehicle Routing Problem

Several automotive companies such as Citroen, Honda, or Toyota have started producing and selling the hybrid electric vehicles (HEVs). HEVs can be powered from batteries or by other distinct types of energy sources, commonly diesel engines. On the one hand, HEVs represent a good trade-off between minimizing energy consumption and polluting emissions, on the other one they have a limited continuous driving range. However, they represent a new significant challenge for environmental vehicles production, and they are an emerging alternative to internal combustion engine vehicles. Doppstadt et al. [25] introduced the HEV traveling salesman problem (HEV-TSP) which is an extension of the well-known traveling salesman problem. The use of different modes of operating of the vehicles causes different costs and driving times for each arc, increasing the complexity of the problem. The authors formulated the HEV-TSP with four working modes: combustion-only, electric-only, charging, and boost modes. Then, they showed the positive effects of using HEVs and solved the larger instances by a tabu search algorithm. Zhen et al. [99] extended this work introducing the HEV routing problem (HE-VRP) with mode selection. As in Doppstadt et al. [25], HEVs have four modes of propulsion, and they considered the mode selection for each road segment. They proposed a swarm optimization algorithm. The plug-in hybrid electric vehicles (PHEVs) are particular HEVs that have only two propulsion modes: an internal combustion engine and an electric engine, that can be easily switched. Vincent et al. [38] introduced the hybrid vehicle routing problem as an extension of the G-VRP.

They focus on the use of Plug-in hybrid vehicles (PHVs) and mild-HEVs that do not have the electric-only propulsion mode. They modeled the problem and proposed an effective simulated annealing to solve it; then, they conducted a sensitivity analysis to understand the effect of HEVs and charging stations on the costs. A similar problem is studied by Mancini [65]. Each HEV has a limited capacity battery and can be charged at a charging station; however, it can change propulsion mode any time. After modeling the problem, the author proposed a large neighborhood search based matheuristic.

1.4.4   Mixed Fleet Green-Vehicle Routing Problem

Gonçalves et al. [37] studied a VRP variant with a mixed fleet composed of EVs and conventional vehicles and pickup and delivery. EVs have a fixed autonomy and charging time, and they can be charged at any time during a route. They applied their model to a particular case of a Portuguese battery distributor and studied three different scenarios: in the first one the fleet is composed only of company's conventional vehicles, in the second one they considered a fleet composed of conventional vehicles and uncapacitated EVs, while in the third one they use only EVs. After solving the model with CPLEX and testing the three scenarios, the authors conclude that the use of EVs leads to a significant cost increase due to the initial investment. Sassi et al. [80] proposed the heterogeneous electric vehicle routing problem with time-dependent charging costs and a mixed fleet composed of conventional vehicles and EVs. The EVs are characterized by different battery capacities and operating costs. An EV can be partially charged only at either the available and compatible CSs or the depot. Charging costs vary according to the station and the time of day. The authors solved the problem with a construction heuristic, as well as an inject-eject routine-based local search. Also Goeke and Schneider [36] formulated the E-VRP with time windows and mixed fleet of conventional vehicles and EVs. EVs can be fully charged at the available CSs. Charging times vary according to the battery level when the EV arrives at the CS. The authors proposed a comprehensive energy consumption model which considers speed, vehicle mass, and gradient. They considered three objective functions: minimize traveled distance, minimize energy and labor costs, minimize cost for battery replacement. To solve the problem, they proposed an ALNS heuristic, which is highly effective also for VRPTW and E-VRPTW on benchmark instances. Yavuz and Çapar [93] modeled a G-VRP with a mixed fleet of AFVs and ICCVs. They studied the benefits of introducing the AFVs in a fleet of ICCVs, considering four different objective functions, i.e., minimization of total miles traveled, total emissions, total costs and ICCVs traveled miles. In their formulation, the authors considered the possibility to recharge AFVs at customer locations. Their study pointed out that the classical vehicle routing objective of minimizing total vehicle miles traveled does not work well for this class of problem; instead, the minimization of emissions leads to more suitable solutions. A VNS heuristic is defined to solve the proposed model. Yavuz [92] extended the work of Yavuz and Çapar [93] by introducing the possibility to recharge AFVs to the external CSs. The author proposed an iterated beam search algorithm based on arc duplication, a parametric algorithm that can work as either an exact or a heuristic method. Koyuncu and Yavuz [49] presented a unified modeling framework comprising node- and arc-duplicating formulations. The models are capable of handling various recharging policies and homogeneous or mixed fleets. They carried out several computational tests and demonstrate that the arc-duplicating formulation outperforms the node-duplicating one. Breunig et al. [12] introduced the electric two-echelon VRP, in which conventional vehicles transport the commodity from the depot to the satellites (first-level), and electrical vehicles serve all the customers starting from the satellites (second-level). They developed a large neighborhood search as well as an exact mathematical programming algorithm to solve that problem. Macrina et al. [63] proposed an ILS algorithm to solve a mixed fleet G-VRP with partial battery recharging and time windows. They imposed a limit on the pollution emissions, evaluated as function of weight, type of fuel, and traveled distance. Macrina et al. [62] introduced a realistic energy consumption model to the G-VRP with mixed fleet. In particular, for each arc they considered three phases: the first phase, the "acceleration phase", in which the value of the acceleration is positive, the second phase named "constant speed phase" where the speed is constant, and the third phase, the "deceleration phase", in which the value of the acceleration is negative. Thus, they computed the instantaneous energy consumption and braking energy regeneration for the EVs. They modeled the problem and solved it with a large neighborhood search.

Hiermann et al. [40] introduced the G-VRP with a mixed fleet of conventional vehicles, PHEVs and EVs, time windows, and charging stations. To solve this problem the authors proposed a hybrid genetic algorithm. Moreover, they showed the benefits of using PHEVs in a fleet composed of conventional vehicles and EVs.

1.4.5 Electric Vehicle Location Routing Problem

The location and the selection of technology of the CSs are crucial in EVs routing. The costs related to the installation of the CSs and the network infrastructure highly impact on the companies decisions. Yang and Sun [91] introduced the electric vehicles battery swap stations location routing problem. The aim is to determine the locations of battery swap stations (SSs), as well as the routing plan of EVs. The authors propose two heuristics for the problem: the first one, named SIGALNS, is a four-phase heuristic including a modified sweep heuristic, iterated greedy algorithm, and ALNS; the second one, named TS-MCWS, is a hybrid heuristic which combines tabu search and Clarke-Wright savings method. Li-ying and Yuan-bin [58] introduced the EV multiple charging station location routing problem with time windows. The authors considered the possibility of choosing among several types of charging infrastructures. They proposed a hybrid heuristic which incorporates an

adaptive VNS with a tabu search algorithm. Schiffer and Walther [81] proposed the
electric location routing problem with time windows and partial recharging. EVs
can be charged at any node in the network and only one type of technology was
considered. The authors modeled three objective functions: minimize total traveled
distance, minimize number of used EVs, minimize the number of CSs. Schiffer
and Walther [82] introduced the location routing problem with intra-route facilities
whose aim is to find the location of facilities for intermediate stops. The facilities
do not necessarily coincide with customers and are not depots. Intra-route facilities
allow for intermediate stops on a route in order to keep the vehicle operational.
The authors proposed an ALNS heuristic including dynamic programming. Paz
et al. [71] proposed the multi-depot electric vehicle location routing problem with
time windows in which a homogeneous fleet of EVs is considered. The goal is to
determine the number and location of CSs and depots, as well as the number of
EVs and their routes. The authors also considered the possibility to charge the EV
to the CSs or to swap the battery to the battery swap stations. Since they considered
different charging strategies, they proposed and tested three models: in the first
one the conventional partial or complete charges can be done at either the depots
or the customer locations, in the second one the batteries can be swapped only
at the depots, while in the third one if a charging vertex is activated, then it is a
battery swap station and a customer vertex is activated only for the conventional
recharging. Zhang et al. [95] studied the electric vehicle location routing problem
considering swap battery stations and stochastic demands. The aim is to minimize
the cost related to the optimal number and location of stations and to the route plan,
based on stochastic customers demands. After modeling the problem, the authors
proposed a hybrid VNS algorithm.

1.4.6   Electric Vehicle Routing Problem with Non-linear Charging Function

All early E-VRP models assumed that the battery charge level is a linear function
of charging time. Since in reality this function is non-linear, Montoya et al.
[68] extended the E-VRP by considering a non-linear charging function. They
proposed an ILS enhanced with a heuristic concentration for the problem. Then,
they conducted several computational experiments, by comparing their proposed
non-linear charging function to those in the literature, and concluded that a linear
function charging may lead to infeasible or expensive solutions. Froger et al. [34]
proposed two new mixed integer programming models for the problem introduced
by Montoya et al. [68]. The first one replaces the node-based tracking of the time
and the state of charge from Montoya et al. [68] model by an arc-based tracking of
these variables. Computational experiments showed that this alternative formulation
outperforms the oldest one. The second model avoids the replication of CS nodes.
Even in this case, the proposed formulation improves the results of the model

with replication. Moreover, the authors proposed a heuristic algorithm and an exact labelling algorithm to solve the problem, which outperforms the existing approach proposed by Montoya et al. [68]. A similar problem is addressed by Froger et al. [33] which introduced CSs with limited number of chargers (one, two, or three) and an EV may need to wait for charging its battery if all the chargers are busy with other vehicles. Keskin et al. [46] considered the time-dependent queueing times at the recharge stations. The authors assumed Poisson arrivals and a first-in-first-out strategy, soft time windows constraints, and a non-linear charging function. They formulated the problem and proposed a matheuristic, which is a combination of ALNS and the solution of their model. Their computational results showed that the waiting time at the charging stations may increase the total cost.

1.5   New Paradigms in VRP

Even if the use of alternative fuel vehicles is increasing and the reduction of negative externalities is one of the most important issue for many countries, a large number of cars, trucks, and motorbikes continue to travel on the roads, generating congestion and polluting emission. Proposing new efficient and effective solutions for transportation is a crucial challenge for companies which have to face off to more and more demanding customers, but have to take into account environmental aspects. The majority of retailers offers to its customers the possibility to receive their orders in few hours, in the same day or in few days, thus it is important to be fast. However, augmenting the speed leads to augment freight movements, hence to increase congestion, noise, and polluting emissions. Last-mile and same day delivery process are becoming the most challenging activities for the companies; hence, several of them have started to adopt and propose new paradigms in VRP. Unmanned aerial vehicles, or commonly named drones, are an emerging technology which is becoming more and more popular in the last-mile delivery. Drones do not need a human pilot on board and usually are piloted by a remote control. They have less environmental impacts and are faster because they are not affected by congestion on the road networks. In the last years, several companies, such as Amazon, Google, DHL, are testing their drones and are ready to use them daily for their deliveries [3, 70]. Hence, several researchers started to study the delivery problem with drones. Murray and Chu [69] introduced the routing problem combining truck and drone. They presented two traveling salesman problems, in the first one a set of customers can be served by either a drone or a truck, in the second one the customers can be served by a single truck or a fleet of several drones. Other extension was proposed by Agatz et al. [2] and Marinelli et al. [66] which considered different features of the problem. Wang et al. [89] introduced the VRP with drones (VRP-D), in which a fleet of trucks is equipped with drones. In this work only drones can deliver parcels, starting their flight from the trucks. Starting from this works, several extensions have been proposed for the VRP-D by several authors (see Poikonen [74], Di Puglia Pugliese and Guerriero [23], Ulmer and Thomas

[88]). Even the use of drones has gained a great appealing, several issues have also to be taken into account. In fact, thinking to deliver all the parcels with drones is not possible, due to their limited load capacity and short operating range. The majority of the proposed models supposes to work with a mixed fleet of conventional vehicles and drones. Thus, in order to reduce the number of conventional vehicles, in the future the possibility to use EVs in tandem with drones could be exploited. In addition, since delivery with drones may become widespread over the next few years, issues related to the regulations, more reliable energy consumption models, public safety and privacy, and air congestion have to be taken into account. Another new delivery solution adopted by several big on-line retailers (e.g., Walmart [7], Amazon [10]) is based on "crowd-shipping" concept. Usually, a large number of cars that daily travel on the roads are not completely exploited. Thus, the main idea is to allow deliveries, which usually are performed by the companies, to ordinary people. Archetti et al. [5] introduced the VRP with occasional drivers (VRPOD). In this problem a fleet of conventional vehicles is supported by ordinary people, named occasional drivers (ODs), which make a deviation from their ordinary routes to deliver parcels, for a small compensation. The authors showed the benefit of using ODs for delivering parcels, by using several compensation schemes for ODs. The VRPOD has been extended by Macrina et al. [61], introducing time windows, multiple deliveries for ODs, and split and delivery policy. A VRP with a mixed fleet of conventional vehicle, EVs, and ODs have been addressed by Macrina and Guerriero [60]. The authors have shown the positive impacts of using ODs and EVs on polluting and routing costs reduction. The reader is referred to Buldeo et al. [15] and Arslan et al. [6] for a complete survey on crowd-logistics and crowd-shipping, respectively. Crowd-shipping is an innovative approach for the last-mile deliveries; however, the companies making use of this system have to face off to several issues: the reliability, satisfaction, and compensation of ODs among the others.

1.6   Conclusions

In recent years the interest in green logistics increased, thus several authors started to study the vehicle routing problem under a green perspective. In this work we have given an overview of the green-vehicle routing (G-VRP) variants introduced during 2011–2019. We have studied and classified the G-VRP variants and discussed the proposed solution approaches. We identified two main classes of that problem and we have described the variants and the proposed approaches. From the literature, it is clear that very few researches are devoted to the mixed fleet G-VRP variant and to the VRP with hybrid vehicles. Only four contributions take into account the non-linear charging function for the electric vehicles. Moreover, no one has combined an energy consumption model considering acceleration and deceleration effects with non-linear charging function. Hence, future researches may focus on these aspects. We have also discussed about two new paradigms in the last-mile delivery, which are drones and crowd-shipping, giving some idea for future researches in these fields.

References

1. Affi, H., Derbel, M., Jarboui, B.: Variable neighborhood search algorithm for the green vehicle routing problem. Int. J. Ind. Eng. Comput. **9**, 195–204 (2018)
2. Agatz, N., Bouman, P., Schmidt, M.: Optimization approaches for the traveling salesman problem with drone. Transp. Sci. **52**(4), 739–1034 (2018)
3. Amazon prime air.   https://www.amazon.com//Amazon-Prime-Air/b?ie=UTF8&node= 8037720011. Accessed 12 Mar 2019
4. Andelmin, J., Bartolini, E.: A multi-start local search heuristic for the green vehicle routing problem based on a multigraph reformulation. Comput. Oper. Res. **109**, 43–63 (2019)
5. Archetti, C., Savelsbergh, M., Speranza, M.G.: The vehicle routing problem with occasional drivers. Eur. J. Oper. Res. **254**(2), 472–480 (2016)
6. Arslan, A.M., Agatz, N., Kroon, L., Zuidwijk, R.: Crowdsourced delivery: a dynamic pickup and delivery problem with ad-hoc drivers. Technical report, ERIM, Report Series Reference (2016)
7. Barr, A., Wohl, J.: Exclusive: Walmart may get customers to deliver packages to online buyers. REUTERS – Business Week (2013)
8. Basso, R., Kulcsár, B., Egardt, B., Lindroth, P., Sanchez-Diaz, I.: Energy consumption estimation integrated into the electric vehicle routing problem. Transp. Res. D Transp. Environ. **69**, 141–167 (2019)
9. Bektaş, T., Laporte, G.: The pollution-routing problem. Transp. Res. B **45**, 1232–1250 (2011)
10. Bensinger, G.: Amazon's next delivery drone: You. Wall Street J. (2015). https://www.wsj. com/articles/amazon-seeks-help-with-deliveries-1434466857
11. Bravo, M., Rojas, L.P., Parada, V.: An evolutionary algorithm for the multi-objective pick-up and delivery pollution-routing problem. Int. Trans. Oper. Res. **26**, 302–317 (2017)
12. Breunig, U., Baldacci, R., Hartl, R.F., Vidal, T.: The electric two-echelon vehicle routing problem. Comput. Oper. Res. **103**, 198–210 (2019)
13. Bruglieri, M., Mancini, S., Pezzella, F., Pisacane, O.: A path-based solution approach for the green vehicle routing problem. Comput. Oper. Res. **103**, 109–122 (2019)
14. Bruglieri, M., Mancini, S.S., Pisacane, O.: More efficient formulations and valid inequalities for the green vehicle routing problem. Transp. Res. C **105**, 283–296 (2019)
15. Buldeo Rai, H., Verlinde, S., Merckx, J., Macharis, C.: Crowd logistics: an opportunity for more sustainable urban freight transport? Eur. Trans. Res. Rev. **9**(3), 39 (2017)
16. Conrad, R.G., Figliozzi, M.A.: The recharging vehicle routing problem. In: Doolen, T., Van Aken, E. (Eds.) Industrial Engineering Research Conference, Reno, Nevada (2011)
17. Costa, L., Lust, T., Kramer, R., Subramanian, A.: A two-phase pareto local search heuristic for the bi-objective pollution-routing problem. Networks **72**, 311–336 (2018)
18. Dantzig, G.B., Ramser, J.H.: The truck dispatching problem. Manag. Sci. **6**(1), 80–91 (1959)
19. de Oliveira da Costa, P.R., Mauceri, S., Carroll, P., Pallonetto, F.: A genetic algorithm for a green vehicle routing problem. Electron. Notes Discrete Math. **64**, 65–74 (2018)
20. Demir, E., Bektaş, T., Laporte, G.: An adaptive large neighborhood search heuristic for the pollution-routing problem. Eur. J. Oper. Res. **223**, 346–359 (2012)
21. Demir, E., Bektaş, T., Laporte, G.: The bi-objective pollution-routing problem. Eur. J. Oper. Res. **232**, 464–478 (2014)
22. Desaulniers, G., Errico, F., Irnich, S., Schneider, M.: Exact algorithms for electric vehicle-routing problems with time windows. Oper. Res. **64**, 1388–1405 (2016)
23. Di Puglia Pugliese, L., Guerriero, F.: Last-mile deliveries by using drones and classical vehicles. In: Sforza, A., Sterle, C. (eds.) International Conference on Optimization and Decision Science, ODS 2017. Springer Proceedings in Mathematics and Statistics, pp. 557–565, Springer New York LLC, New York (2017)
24. Ding, N., Battay, R., Kwon, C.: Conflict-free electric vehicle routing problem with capacitated charging stations and partial recharge (2015). https://www.chkwon.net/papers

25. Doppstadt, C., Koberstein, A., Vigo, D.: The hybrid electric vehicle – traveling salesman problem. Eur. J. Oper. Res. **253**(3), 825–842 (2016)
26. Dukkanci, O., Kara, B.Y., Bektaş, T.: The green location-routing problem. Comput. Oper. Res. **105**, 187–202 (2019)
27. Ehmke, J.F., Campbell, A.M., Thomas, B.W.: Vehicle routing to minimize time-dependent emissions in urban areas. Eur. J. Oper. Res. **251**(2), 478–494 (2016)
28. Erdelić, T., Carić, T.: A survey on the electric vehicle routing problem: Variants and solution approaches. J. Adv. Transp. **2019**, 48 (2019). https://doi.org/10.1155/2019/5075671
29. Erdoğan, S., Miller-Hooks, E.: A green vehicle routing problem. Transp. Res. E **48**(1), 100–114 (2012)
30. Felipe, A., Ortuño, M.T., Righini, G., Tirado, G.: A heuristic approach for the green vehicle routing problem with multiple technologies and partial recharges. Transp. Res. E **71**, 111–128 (2014)
31. Figliozzi, M.A.: The impacts of congestion on time-definitive urban freight distribution networks CO_2 emission levels: Results from a case study in Portland, Oregon. Transp. Res. C **19**(5), 766–778 (2011)
32. Franceschetti, A., Honhon, D., Van Woensel, T., Bektaş, T., Laporte, G.: The time-dependent pollution-routing problem. Transp. Res. B **56**, 265–293 (2013)
33. Froger, A., Mendoza, J.E., Jabali, O., Laporte, G.: Matheuristic for the electric vehicle routing problem with capacitated charging stations. Technical report (2017). https://hal.archives-ouvertes.fr/hal-01559524/document
34. Froger, A., Mendoza, J.E., Jabali, O., Laporte, G.: Improved formulations and algorithmic components for the electric vehicle routing problem with nonlinear charging functions. Comput. Oper. Res. **104**, 256–294 (2019)
35. Goeke, D.: Granular tabu search for the pickup and delivery problem with time windows and electric vehicles. Eur. J. Oper. Res. **278**, 821–836 (2019)
36. Goeke, D., Schneider, M.: Routing a mixed fleet of electric and conventional vehicles. Eur. J. Oper. Res. **245**, 81–99 (2015)
37. Gonçalves, F., Cardoso, S.R., Relvas, S., Barbosa-Póvoa, A.P.F.D.: Optimization of a distribution network using electric vehicles: A VRP problem. In: 15th Congresso Nacional da Associação Portuguesa de Investigação Operacional, pp. 18–20 (2011)
38. Hidayat, Y.A., Vincent, F.Y., Redi, A.A.N.P., Wibowo, O.J.: A simulated annealing heuristic for the hybrid vehicle routing problem. Appl. Soft Comput. **53**, 119–132 (2017)
39. Hiermann, G., Puchinger, J., Ropke, S., Hartl, R.F.: The electric fleet size and mix vehicle routing problem with time windows and recharging stations. Eur. J. Oper. Res. **252**, 995–1018 (2016)
40. Hiermann, G., Hartl, J., Puchinger, R.F., Vidal T.: Routing a mix of conventional, plug-in hybrid, and electric vehicles. Eur. J. Oper. Res. **272**, 235–248 (2019)
41. Hooshmand, F., MirHassani, S.A.: Time dependent green VRP with alternative fuel powered vehicles. Energy Syst. **10**, 721–756 (2019)
42. Jabali, O., Van Woensel, T., de Kok, A.G.: Analysis of travel times and CO_2 emissions in time-dependent vehicle routing. Prod. Oper. Manag. **21**(6), 1060–1074 (2012)
43. Joo, H., Lim, Y.: Ant colony optimized routing strategy for electric vehicles. J. Adv. Transp. **2018**, 9 (2018)
44. Kancharla, S., Ramadurai, G.: Incorporating driving cycle based fuel consumption estimation in green vehicle routing problems. Sustain. Cities Soc. **40**, 214–221 (2018)
45. Keskin, M., Çatay, B.: Partial recharge strategies for the electric vehicle routing problem with time windows. Transp. Res. C **65**, 111–127 (2016)
46. Keskin, M., Laporte, G., Çatay, B.: Electric vehicle routing problem with time-dependent waiting times at recharging stations. Comput. Oper. Res. **107**, 77–94 (2019)
47. Koç, Ç., Karaoglan, I.: The green vehicle routing problem: A heuristic based exact solution approach. Appl. Soft Comput. **39**, 154–164 (2016)
48. Koç, Ç, Bektaş, T., Jabali, O., Laporte, G.: The fleet size and mix pollution-routing problem. Transp. Res. B **70**, 239–254 (2014)

49. Koyuncu, I., Yavuz, M.: Duplicating nodes or arcs in green vehicle routing: a computational comparison of two formulations. Transp. Res. E **122**, 605–623 (2019)
50. Kramer, R., Maculan, N., Subramanian, A., Vidal, T.: A speed and departure time optimization algorithm for the pollution-routing problem. Eur. J. Oper. Res. **247**, 782–787 (2015)
51. Kramer, R., Subramanian, A., Vidal, T., Cabral, L.A.F.: A matheuristic approach for the pollution-routing problem. Eur. J. Oper. Res. **243**, 523–539 (2015)
52. Kumar, N.S., Paneerselvam, R.: A survey on the vehicle routing problem and its variants. Intell. Inf. Manag. **4**, 66–74 (2012)
53. Laporte, G.: The vehicle routing problem: An overview of exact and approximate algorithms. Eur. J. Oper. Res. **59**(3), 345–358 (1992)
54. Laporte, G.: What you should know about the vehicle routing problem. Naval Res. Logist. **54**(8), 811–819 (2007)
55. Leggieri, V., Haouari, M.: A practical solution approach for the green vehicle routing problem. Transp. Res. E **104**, 97–112 (2017)
56. Lin, J., Zhou, W., Wolfson, O.: Electric vehicle routing problem. In: Transportation Research Procedia, pp. 508–521, Tenerife, Canary Islands (Spain), June 17–19 (2009). The 9th International Conference on City Logistics
57. Lin, C., Choy, K.L., Ho, G.T.S., Chung, S.H., Lam, H.Y.: Survey of green vehicle routing problem: Past and future trends. Expert Syst. Appl. **41**(4), 1118–1138 (2014)
58. Li-ying, W., Yuan-bin, S.: Multiple charging station location-routing problem with time window of electric vehicle. J. Eng. Sci. Technol. Rev. **8**(5), 190–201 (2015)
59. Li-ying, W., Yuan-bin, S.: A variable neighborhood search branching for the electric vehicle routing problem with time windows. Electron. Notes Discrete Math. **47**, 221–228 (2015)
60. Macrina, G., Guerriero, F.: The green vehicle routing problem with occasional drivers. In: Daniele, P., Scrimali, L. (eds.) New Trends in Emerging Complex Real Life Problems. Springer International Publishing, Springer New York LLC, New York (2018)
61. Macrina, G., Di Puglia Pugliese, L., Guerriero, F., Laganà, D.: The vehicle routing problem with occasional drivers and time windows. In: Sforza, A., Sterle, C. (eds.) Optimization and Decision Science: Methodologies and Applications. Springer Proceedings in Mathematics Statistics, Cham. ODS, Sorrento, vol. 217, pp. 577–587. Springer, Cham (2017)
62. Macrina, G., Laporte, G., Guerriero, F., Di Puglia Pugliese, L.: An energy-efficient green-vehicle routing problem with mixed vehicle fleet, partial battery recharging and time windows. Eur. J. Oper. Res. **276**(3), 971–982 (2019)
63. Macrina, G., Di Puglia Pugliese, L., Guerriero, F., Laporte, G.: The green mixed fleet vehicle routing problem with partial battery recharging and time windows. Comput. Oper. Res. **101**, 183–199 (2019)
64. Majidi, S., Hosseini-Motlagh, S.M., Ignatius, J.: Adaptive large neighborhood search heuristic for pollution-routing problem with simultaneous pickup and delivery. Soft Comput. **22**, 2851–2865 (2018)
65. Mancini, S.: The hybrid vehicle routing problem. Transp. Res. C Emerg. Technol. **78**, 1–12 (2017)
66. Marinelli, M., Caggiani, L., Ottomanelli, M., Dell'Orco, M.: En route truck–drone parcel delivery for optimal vehicle routing strategies. IET Intell. Transp. Syst. **12**(4), 253–261 (2017)
67. Montoya, A., Guéret, C., Mendoza, J.E., Villegas, J.G.: A multi-space sampling heuristic for the green vehicle routing problem. Transp. Res. C **70**, 113–128 (2016)
68. Montoya, A., Guéret, C., Mendoza, J.E., Villegas, J.G.: The electric vehicle routing problem with nonlinear charging function. Transp. Res. B **103**, 87–110 (2017)
69. Murray, C.C., Chu, A.G.: The flying sidekick traveling salesman problem: Optimization of drone-assisted parcel delivery. Transp. Res. C Emerg. Technol. **54**, 86–109 (2015)
70. Parcelcopter: DHL's drone. https://discover.dhl.com/business/business-ethics/parcelcopter-drone-technology. Accessed 12 Mar 2019
71. Paz, J.C., Granada-Echeverri, M., Escobar, J.W.: The multi-depot electric vehicle location routing problem with time windows. Int. J. Ind. Eng. Comput. **9**, 123–136 (2018)

72. Pelletier, S., Jabali, O., Laporte, G.: Goods distribution with electric vehicles: Review and research perspectives. Transp. Sci. **50**(1), 3–22 (2016)
73. Pelletier, S., Jabali, O., Laporte, G.: The electric vehicle routing problem with energy consumption uncertainty. Transp. Res. B **126**, 225–255 (2019)
74. Poikonen, S., Wang, X., Golden, B.: The vehicle routing problem with drones: Extended models and connections. Networks **70**(1), 34–43 (2017)
75. Poonthalir, G., Nadarajan, R.: A fuel efficient green vehicle routing problem with varying speed constraint (F-GVRP). Expert Syst. Appl. **100**, 131–144 (2018)
76. Psaraftis, H.N.: Green transportation in logistics: The quest for win-win solutions. In: International Series in Operations Research & Management Science, vol. 226. Springer International Publishing (2016)
77. Qian, J., Eglese, R.: Fuel emissions optimization in vehicle routing problems with time-varying speeds. Eur. J. Oper. Res. **248**, 840–848 (2016)
78. Raeesi, R., Zografos, K.G.: The multi-objective Steiner pollution-routing problem on congested urban road networks. Transp. Res. B **122**, 457–485 (2019)
79. Rauniyar, A., Nath, R., Muhuri, P.K.: Multi-factorial evolutionary algorithm based novel solution approach for multi-objective pollution-routing problem. Comput. Ind. Eng. **130**, 757–771 (2019)
80. Sassi, O., Cherif, W.R., Oulamara, A.: Vehicle routing problem with mixed feet of conventional and heterogenous electric vehicles and time dependent charging costs. Technical report (2014). https://hal.archives-ouvertes.fr/hal-01083966
81. Schiffer, M., Walther, G.: The electric location routing problem with time windows and partial recharging. Eur. J. Oper. Res. **260**, 995–1013 (2017)
82. Schiffer, M., Walther, G.: An adaptive large neighborhood search for the location routing problem with intra-route facilities. Transp. Sci. **52**, 229–496 (2018)
83. Schneider, M., Stenger, A., Goeke, A.: The electric vehicle routing problem with time windows and recharging stations. Transp. Sci. **48**(4), 500–520 (2014)
84. Shao, S., Guan, W., Ran, B., He, Z., Bi, Z.: Electric vehicle routing problem with charging time and variable travel time. Math. Probl. Eng. **2017**, 13 (2017)
85. Suzuki, Y.: A dual-objective metaheuristic approach to solve practical pollution routing problem. Int. J. Prod. Econ. **176**, 143–153 (2016)
86. Tajik, N., Tavakkoli-Moghaddama, R., Vahdani, B., Meysam Mousavic, S.: A robust optimization approach for pollution routing problem with pickup and delivery under uncertainty. J. Manuf. Syst. **33**, 277–286 (2014)
87. Toro, E.M., Franco, J.F.: A multi-objective model for the green capacitated location-routing problem considering environmental impact. Comput. Ind. Eng. **11**, 114–125 (2017)
88. Ulmer, M.W., Thomas, B.W.: Same-day delivery with a heterogeneous fleet of drones and vehicles. Technical report, Technical University of Braunschweig (2017)
89. Wang, X., Poikonen, S., Golden, B.: The vehicle routing problem with drones: several worst-case results. Optim. Lett. **11**(4), 679 (2017)
90. Xiao, Y., Konak, A.: The heterogeneous green vehicle routing and scheduling problem with time-varying traffic congestion. Transp. Res. E **88**, 146–166 (2016)
91. Yang, J., Sun, H.: Battery swap station location-routing problem with capacitated electric vehicles. Comput. Oper. Res. **55**, 217–232 (2015)
92. Yavuz, M.: An iterated beam search algorithm for the green vehicle routing problem. Networks **69**(3), 317–328 (2017)
93. Yavuz, M., Çapar, I.: Alternative-fuel vehicle adoption in service fleets: impact evaluation through optimization modeling. Transp. Sci. **51**, 480–493 (2017)
94. Yu, Y., Wang, S., Wang, J., Huang, M.: A branch-and-price algorithm for the heterogeneous fleet green vehicle routing problem with time windows. Transp. Res. B **122**, 511–527 (2019)
95. Zhang, S., Chen, M., Zhang, W.: A novel location-routing problem in electric vehicle transportation with stochastic demands. J. Clean. Prod. **221**, 567–581 (2019)
96. Zhang, S., Zhang, W., Gajpal, Y., Appadoo, S.S.: Ant colony algorithm for routing alternate fuel vehicles in multi-depot vehicle routing problem. Decision Science in Action: Theory and

Applications of Modern Decision Analytic Opimization, pp. 251–260. Springer, Singapore (2019)
97. Zhang, S., Gajpal, Y., Appadoo, S.S., Abdulkader, M.M.S.: Electric vehicle routing problem with recharging stations for minimizing energy consumption. Int. J. Prod. Econ. **203**, 404–413 (2018)
98. Zhao, L., Van Woensel, T., Gross, J.P., Huang, Y.: Time-dependent vehicle routing problem with path flexibility. Transp. Res. B **95**, 169–195 (2017)
99. Zhen, L., Xu, Z., Ma, C., Xiao, L.: Hybrid electric vehicle routing problem with mode selection. J. Prod. Res. (2019). https://doi.org/10.1080/00207543.2019.1598593

Chapter 2
An Integrated Location-Inventory Routing Problem for ATMs in Banking Industry: A Green Approach

Nader Nazari-Ganje and S. Mohammad J. Mirzapour Al-E Hashem

Abstract In today's competitive banking industry, deciding on the launch of new facilities is very important, and the lack of attention to this issue will lead to heavy costs. Automated Teller Machine (ATM) Branch is one of the most important facilities in the banking industry. In the current competitive market, choosing the optimal location for these facilities beside an optimal weekly routing for their cash replenishment is very important from the economic and environmental aspects. In this study, a novel mathematical model is presented to integrate the two well-studied location and vehicle routing problems for ATMs in banking industry. In location side, considering the cost of deployment, the model tries to find the optimal new ATMs location to maximize the coverage of ATM branches and the bank's share in competing with branches of other existing banks by using the different elements of a gravity function. Then, by aware of the location of the new facilities, the model concurrently attempts to provide an optimal cash replenishment policy by embedding a green vehicle routing problem and taking into account a central warehouse and several types of banknote, in order to minimize total costs including the transportation, disposition and shortage costs such that the total GHG emissions generated by the cash carrier vehicles are also minimized. The proposed model is applied in a real case study and the results show that the model can be effectively used by the bankers to increase the performance of the ATMs network and pragmatically contribute to social efforts in response to the environmental concerns.

N. Nazari-Ganje · S. M. J. Mirzapour Al-E Hashem (✉)
Department of Industrial Engineering and Management Systems, Amirkabir University
of Technology, (Tehran Polytechnic), Tehran, Iran
e-mail: mirzapour@aut.ac.ir

© Springer Nature Switzerland AG 2020
H. Derbel et al. (eds.), *Modeling and Optimization in Green Logistics*,
https://doi.org/10.1007/978-3-030-45308-4_2

2.1 Introduction

With the highest growth among all electronic hardware, ATMs are, nowadays, considered to be the most popular one in the banking industry. Market studies in different countries have shown that most bank customers have preferred using ATMs to going to the bank. Iran Central Bank statistics shows that although the number of bank cards issued by the end of July 2017 has exceeded 300 million, ATMs have not yet developed in proportion to this number and need to grow proportionately. Statistics (by the end of July 2018) shows that 45,000 ATMs have been in use in Iran which means only one ATM for every 8279 bank cards. This has caused long queues at such specific times as cash subsidy paydays, end-of-month paydays, school year commencement (September and early October), or new year commencement (March). Selecting right locations for ATMs will highly improve their acceptance by customers and will have many monetary/non-monetary benefits for banks. For instance, banks can expand their activities, serve customers better, increase their profitability, and reduce their service costs; however, since ATMs involve high costs, their improper locating decisions will cause banks to incur high financial losses. The more appropriate is a bank location, the more is its accessibility; hence, the lower is its access costs for the customers. As regards the liquidity management and operational costs, serving ATMs is a costly task, and since the cost price is high in developing countries, making the performance efficiency important, most banks have focused on how to manage the cash money in ATMs which means how much money they should keep in the machine to avoid liquidity excess/shortage. The optimum liquidity management and service accessibility are important factors in the ATM network service business because they can help banks to manage their systems dynamically; hence, recently, most banks are shifting their attention towards gaining higher efficiency in managing the liquidity in ATMs. The issue is important because if the cash replenishment is less than the actual demand, customers will be dissatisfied, and if it is more, the bank will incur extra costs. Excess cash is not economical in an ATM, but enough cash will increase the service level and lessen the maintenance costs creating a trade-off.

Although a proper cash replenishment, as previously mentioned, has a positive impact on both market share and customer satisfaction, but it imposes a significant cost to the system because of the related logistics and transportation activities. Transportation on the other hand, is the main recognized contributor to the CO_2 concentration in the atmosphere, itself is the main factor of global warming. The cash replenishment therefore should be investigated from the both economic and environmental aspects, concurrently.

This paper presents two approaches: hierarchical and integrated; in the former, the optimal bank location is selected first and then its replenishment is planned, but in the latter, the replenishment costs are simultaneously considered when selecting the optimal location. The locating model is to maximize the ATM share of a known bank in a region where other banks compete too so as to maximize its coverage. The first model has two objective functions; the first one attempts to maximize the ATMs' attraction function to set up new branches by considering inter-bank competitions and the second maximizes their coverage. The second model is aimed to minimize total cost of ATMs' cash replenishment including the cash deposition, deficiency, transportation, and the related carbon emission. The paper objective is to find out in which approach (hierarchical or integrated) the decisions are more appropriate. Indeed, it is well known that the integrated approaches provide better decisions than the hierarchical ones. However, in the computational section a sensitivity analysis on the evaluation error of the actual demand is provided to show that on which level of estimation accuracy, the integrated approach still remains dominant. Since in the integrated approach the demand for the ATMs should be predicted prior to locating the ATMs, while in hierarchical approach, the demands are forecasted based on the historical data, since the ATMs are already located. The rest of this paper has been so organized as to review the literature on the ATM locating in Sect. 2.2, describe the problem and present the nonlinear mathematical model in Sect. 2.3, solve the model and perform sensitivity analyses on the parameters and compare the two hierarchical approach and the integrated model in Sect. 2.4, and present the conclusions and suggestions for future studies in Sect. 2.5.

2.2   Literature Review

The numerous studies done, so far, on ATMs are of five categories: (1) those introducing the ATM receptors' personal characteristics (age, sex, education, income level, etc.) [1–4], (2) those introducing important factors that encourage banks to invest and increase the number of ATMs (reduced costs and increased profit, customer satisfaction, market share, and areas where the bank can cover and always be present) [5], (3) those examining the customers' understanding of the service quality (increased accountability, security, reliability, and accessibility, and online banking transactions) [6, 7], (4) those addressing geographical ATM locations [8–18], and (5) those addressing the prediction and optimization of the ATMs' cash values [19–21].

2.2.1   ATM Location Literature Review

Miliotis et al. [8] solved the problem by using Maximal Covering Location Model (MCLM). They mixed MLCM with Geographical Information System (GIS); the basis of the recommended models is to utilize GIS to represent different banking service requests, diverse priorities, and competition in each region. Mentioned information is considered by request satisfaction models which trying to satisfy array of requests and finally, the proposed approach is applied in a major Greek commercial bank. Aldajani et al. [9] performed a convolution-based ATM locating study to determine their minimum number with maximum customer demand coverage in a known geographical area. Li et al. [10] addressed the ATM locating by a PSO (particle swarm optimization) model, studied the planning parameters and ATM locating models, and did decision-making and predictions by combining the geographic information with mathematical models. Alhaffa et al. [11] used Aldajani's heuristic algorithm with three different techniques: (1) Heuristic Algorithm based on Convolution (HAC), (2) Ranked Genetic Algorithm based on Convolution (RGAC), and (3) Simulated Annealing based on Convolution (SAC), and, finally, made a detailed comparison between these methods by doing three experiments. Angelos [12] proposed a geographic information system to evaluate the optimal facility location in a hierarchical network and solved the problem by Voronoi Diagrams to define overflow areas and minimize the average distance for users. Mourad et al. [13] studied the ATM location/allocation optimization by identifying and selecting a number of important bank customer locating criteria based on which they developed an analytical model for locating ATMs. Their model considered the interests of both customers and banks together and was aimed to determine the optimal number of ATMs to cover a specific geographic area. Using a combination of the theoretical studies and algorithmic methods, Moradi et al. [14] proposed a heuristic ATM locating model in the 2D space based on such criteria as age, income, education, and such other considerations as proportional geographical service distribution, areas' available capacity utilization, potential customers' increased access, and so on considering minimum resources and related global standards. Ehsani et al. [15] proposed a mathematical model for locating branches and ATMs aiming at minimizing the facility setup costs with such constraints as the population, facility accessibility, and so on. Awaghade et al. [16] performed site selection and analysis of the nearest available location for ATMs as a case study in India and tried to identify the potential ATM locations in Aundh district, Pune city, using the GIS and RS (remote sensing) methods. Bilginol et al. [17] used the conventional least squares error method to select the optimum ATM locations by first analyzing all the important criteria by the correlation method, then examining the effective ones by the regression model of the least squares error, and, finally, presenting the optimum locations for branch setups. Trang et al. [18] studied the effects of the ATM location on its use, checked the location-use relationship by the Theory of Planned Behavior (TPB) model, and showed, through the results of 398 ATM studies, that the two were positively correlated.

Among studies on the optimum ATM location (regarding only meeting the demand), none has considered the competitors present in the market (Table 2.1). This issue is worth examining from two attraction perspectives: (1) that of the bank that seeks to reduce the setup costs and increase the maximum coverage and (2) that of the customers' (referred to, in this paper, as the intrinsic attraction of the candidate points from the customers' points of view); it is the first time the partial coverage function is discussed in the literature.

Table 2.1 ATM locating literature review

	GIS	Multi-objective	Competitive	Coverage	Partial coverage	Gravity models	Math model	Solution	Replenishment
Miliotis [8]	✓	–	✓	✓	–	–	✓	GIS	–
Aldajani [9]	✓	–	–	✓	✓	–	✓	Convolution	–
Li [10]	✓	–	–	–	–	–	–	PSO	–
Alhaffa [11]	✓	–	–	✓	✓	–	✓	HAC,RGAC,SAC	–
Angelos [12]	✓	–	–	✓	–	–	–	VD,DTS	–
Mourad [13]	✓	AHP	–	✓	–	–	✓	P-median,MCDM	–
Moradi [14]	✓	–	–	✓	✓	–	✓	Convolution,MCDM	–
Ehsani [15]	✓	–	–	✓	–	–	✓	Fuzzy	–
Awaghade [16]	✓	–	–	–	–	–	–	–	–
Bilginol [17]	–	–	–	–	–	–	–	Heuristic	–
Thu Trang [18]	–	–	–	–	–	–	–	TPB	–
This research	✓	✓	✓	✓	✓	✓	✓	CPLEX	✓

GIS (geographical information systems), PSO (Particle Swarm Optimization), HAC (Heuristic Algorithm based on Convolution), RGAC (Ranked Genetic Algorithm based on Convolution), SAC (Simulated Annealing based on Convolution), VD (Voronoi Diagrams), DTS (Directed Tabu Search), AHP (analytic hierarchy process), MCDM (Multi Criteria Decision-Making), TPB (Theory of Planned Behavior)

2.2.2   ATM Replenishment Literature Review

The ATM replenishment literature involves two categories: (1) demand forecasting [19, 20, 22–31] and (2) replenishment planning. Kumar et al. [19] used the K-means clustering to develop a neural network (NN)-based cash money forecasting model. To calculate the error precision, use is made of the mean square error method where the prediction accuracy of each cluster is increased through the use of the NN. Bilir et al. [20] optimized the ATMs' cash values using an integrated cash optimization-required cash prediction model and showed that the integrated model reduced the cash level significantly and increased the customer satisfaction. Kurdel et al. [21] studied the optimal routing to refill ATMs considering the inventory control and vehicle routing costs. The optimum strategy is to focus on reducing cash-related costs while maintaining ATM service levels. Anholt et al. [32] addressed an inventory routing problem considering withdrawals and remittances in the ATMs. After the ATM clustering, they assigned a warehouse to each cluster and refilled each branch through its own cluster warehouse. To solve mixed integer programming (MIP) models, Larrain et al. [33] proposed a novel hybrid algorithm called the variable neighborhood descent MIP (VND MIP) and solved it by the branch and bound algorithm; for large-scale problems, use was made of a heuristic method called the VMND. Aiming at minimizing the maintenance and replenishment costs by savings in refilling multiple ATMs simultaneously, Zhang et al. [34] considered the ATM replenishment policies with the subset costs. There is a Markov decision-making process model that minimizes the average long-term costs.

Studies on the ATMs' cash replenishment are few and none has considered various types of the cash money (Table 2.2). This paper presents, for the first time, an integrated model of ATM locating and replenishment.

2.3   Problem Description

This paper has proposed three models: the first one is related to the ATM locating, the second one deals with the ATM replenishment, and the third one combines the

Table 2.2 ATM Replenishment literature review

	Routing	Multi vehicle	Multi item	Forecasting	Location
[19, 20, 22–31]	–	–	–	✓	–
Kurdel [21]	✓	✓	–	–	–
van Anholt [32]	✓	✓	–	–	–
Larrain [33]	✓	✓	–	–	–
Yu Zhang [34]	✓	✓	–	–	–
This research	✓	✓	✓	–	✓

ATM locating and replenishment. The first two are hierarchical meaning that the locating results are used as inputs to the replenishment model and the third one does the ATM locating and refilling together.

2.3.1   The ATM Locating Model

This model analyzes the customer behavior by the Hoff probabilistic attraction function. The competitive model is static and the locating space is discrete. Besides distance, other factors affecting the customer attraction (e.g. quality of branches and competitors) are also considered in defining the branch attraction. The model objective is to increase the demand attraction and thus maximize the market share for the all own ATM network. The coverage distance is partial and the total coverage radius and maximum coverage radius are specified.

2.3.1.1   Sets

- I: Set of demand points; $i = 1.2. \ldots I$
- J: Set of facility candidate points
- Z: Set of competitor points
- F: Set of attributes defined for the attraction of the candidate points; $f = 1.2. \ldots F$

2.3.1.2   Parameters

- β_f: parameter indicating the relative importance of the fth property
- E_{fj}: Weight of the fth property of branch j
- A_j: Attraction (quality) of branch j
- α: Adjustment coefficient of index A_j
- a_{ij}: partial coverage function
- λ: Adjustment coefficient of index a_{ij}
- ρ_j: Utilization factor of the ATM branch
- d_{ij}: Distance of node i from candidate point j
- S: Distance a facility totally covers
- T: maximum coverage distance
- g: gravity when $nc_j = 0$
- nc_j: Sum of the own and competitor facilities in the coverage area of candidate point j
- $f(nc_j)$: Attraction function of rival bank branches for facility j from customers' points of view
- $Facility$: Maximum number of branches to be set up in candidate locations
- U_{ij}: Attraction function between branch j and customer i

- ε: Minimum coverage created for candidate point i by branch j
- P_{ij}: Percent customers of node i who choose ATM j
- W_i: Demand weight of customer i
- b_j: Setup cost of each candidate point
- $Budget$: Defined budget

2.3.1.3 Decision Variables

- MS_j: Market share provided for facility j
- nb_j: Number of branches set up in the coverage area of branch j
- y_j: Equals 1 if there is ATM in candidate point j and 0 otherwise
- x_{ij}: Demand i met fully or gradually by facility j

2.3.1.4 Modeling

$$max\, Z_1 = \sum_{j \in J} MS_j \times y_j \qquad (2.1)$$

$$max\, Z_2 = \sum_{i \in I} \sum_{j \in J} a_{ij} \times y_j \qquad (2.2)$$

$$MS_j = \frac{\sum_{i \in I} W_i \times P_{ij}}{nc_j + nb_j} \qquad \forall j \in J \qquad (2.3)$$

$$nb_j = \sum_{i \in J \,\&\, i \neq j} a_{ij} \times y_i + 1 \qquad \forall j \in J \qquad (2.4)$$

$$\sum_{j} y_j \leq Facility \qquad (2.5)$$

$$\sum_{j \in J} b_j \times y_j \leq Budget \qquad (2.6)$$

$$x_{ij} \leq y_j \qquad \forall j \in J, i \in I \qquad (2.7)$$

$$x_{ij}, y_j \in \{0, 1\} \ \ 0 \leq MS_j \leq 1, \ 0 \leq nb_j \ \forall j \in J, i \in I \qquad (2.8)$$

Regarding the proposed model, Eq. (2.1) is the objective function of the increased market share, Eq. (2.2) is the objective function of the maximum coverage, Eq. (2.3) shows the market share for own ATMs considering all the branches (own, competitors, and new ones being set up). Equation (2.4) calculates the number of

ATM branches located within the coverage area of branch j, and Constraint (2.5) ensures that the number of facilities may not exceed its upper bound (Facility). Constraint (2.6) shows budget limitations. Constraint (2.7) shows that x_{ij} and y_j are dependent variables, meaning that customer i can be assigned to facility j, if and only if the facility j is set up, and finally, Constraints (2.8) show the variable types.

2.3.1.5   Linearization

$MS_j \times y_j$ in Eq. (2.1) is a bilinear term, since y_j is a binary variable and MS_j is between zero and one, it can be linearized by adding a new auxiliary variable $(0 \leq V_j)$ and adding the following constraints[35]:

$$MS_j \times y_j \longrightarrow V_j$$

$$MS_j - (1 - y_j) \leq V_j \leq MS_j$$

$$MS_j \leq y_j$$

Constraint (2.3) has also an explicit nonlinear term. In order to linearize it, we apply the following arithmetic operations.

$$(1) MS_j = \frac{\sum_{i \in I} W_i \times P_{ij}}{nc_j + nb_j} \qquad \forall j \in J$$

$$(2) nb_j = \sum_{i \in J \& i \neq j} a_{ij} \times y_i + 1 \qquad \forall j \in J$$

$$(1), (2) MS_j = \frac{\sum_{i \in I} W_i \times P_{ij}}{nc_j + \sum_{i \in J \& i \neq j} a_{ij} \times y_i + 1}$$

$$\longrightarrow MS_j \times nc_j + MS_j \times \sum_{i \in J \& i \neq j} a_{ij} \times y_i + MS_j = \sum_{i \in I} W_i \times P_{ij}$$

$$\longrightarrow MS_j \times nc_j + \sum_{i \in J \& i \neq j} a_{ij} \times MS_j \times y_i + MS_j = \sum_{i \in I} W_i \times P_{ij}$$

Then, since y_i is a binary variable and MS_j is bounded by 1, we can replace the bilinear term $MS_j \times y_i$ with a positive auxiliary variable u_{ij}. So, we have the linear equivalent equation as follows:

$$MS_j \times nc_j + \sum_{i \in J \& i \neq j} a_{ij} \times u_{ij} + MS_j = \sum_{i \in I} W_i \times P_{ij}$$

To complete the linearization, the following constraints should be also added:

$$MS_j - (1 - y_i) \le u_{ij} \le MS_j$$

$$u_{ij} \le y_i$$

It is noteworthy that in the other proposed models, we applied the same linearization technique wherever needed.

2.3.1.6 Parameter Calculations

w_i is the demand weight of customer i and its calculation needs the customers (population groups, industrial centers, business centers, etc.) to be identified first. Then, experts assign weight to each customer considering its application, and since the revenue and capital of two similar but different-size customers are different, other normalized weights are assigned to them as the second weight proportional to the area of each application except residential for which the second weight is assigned in proportion to its population. The product of these two weights is the final weight for each customer (see Sect. 2.4).

p_{ij} calculates the probability that customer i may visit branch j and is equal to the attraction function between that customer and the desired branch divided by the sum of all the attraction functions between that customer and all the branches existing in the market.

$$P_{ij} = \frac{U_{ij}}{\sum_j U_{ij}}, \quad \forall j \epsilon J, i \epsilon I \tag{2.9}$$

Based on Hoff model, U_{ij} is the attraction between customer i and candidate point j which is directly proportional to the power of branch j quality and inversely to that of the distance from the customer i to the candidate point j.

$$U_{ij} = A_j^{\alpha} \times a_{ij}^{\lambda} \times \rho_j \times f(nc_j) \tag{2.10}$$

where α and λ are the attraction importance and distance from a branch, respectively, and ρ_j is the branch utilization coefficient which means how much the ATM is utilized in the defined area. Since an increase in the number of bank branches around a candidate point will increase its attraction from the customer's point of view, use has been made of an ascending function, $f(nc_j)$, to include this effect in the model because it increases with an increase in the number of branches.

$$f(nc_j) = ln(nc_j + 1) + g, \qquad \forall j \epsilon J \tag{2.11}$$

According to Eq. (2.11), since an increase in the number of ATMs (own and rivals) will increase the attraction for customers, an ascending function has been defined that will equal g if other branches are none, and will increase to reach a constant value if branches are increased.

$$
a_{ij} = \begin{cases} 1 & d_{ij} \leq S & \forall j \in J i \in I \\ e^{\frac{-(d_{ij}-S)^2}{2 \times (\frac{T-S}{3})^2}} & S < d_{ij} \leq T & \forall j \in J i \in I \\ \varepsilon & d_{ij} > T & \forall j \in J i \in I \end{cases} \tag{2.12}
$$

a_{ij} equals 1 if branch z (or demand point i) is located at a distance S from ATM branch j, but if it is located between T and S, it will be defined as a Gaussian descending function, the peak height of which is 1 and starts decreasing with an increase in d_{ij} until it reaches ε for values greater than T. Therefore, by using Eq. (2.13), nc_j partially calculates the sum of the branches (own and rivals) in the coverage area of candidate point j meaning that branches farther than S have values less than one.

$$
nc_j = \sum_{i \in Z} a_{ij} \qquad \forall j \in J \tag{2.13}
$$

For the calculation of A in Eq. (2.10) which is the attraction or gravity of candidate point j, we should first determine important controllable factors (service, organizational dependence, physical conditions, etc.) that affect customers' attraction, and then, proportional to its importance in attracting a customer, assign a 0–1 weight to each with the help of bank experts and customers. Finally, by using Eq. (2.14) we combine these factors to find the gravity of each ATM branch (Nakanishi and Cooper model [36]).

$$
A_j = \Pi_{f=1}^{n} E_{fj}^{\beta_f} \qquad \forall j \in J \tag{2.14}
$$

where β_f is a parameter to show how important each attraction attribute, we must ensure that sum of the weights of candidate points' properties equals 1.

$$
\sum_{f \in F} \beta_f = 1 \tag{2.15}
$$

2.3.2 The ATM Replenishment Model

This model considers a central warehouse along with a number of ATM branches associated with their predetermined locations and periodic demands, and tries to find the optimum amount of money to be injected to each branch in each time interval by defining two objective functions that minimize the transportation/maintenance/deficiency costs and CO_2 emissions[37].

2.3.2.1 Sets

- I: Set of nodes
- J: Set of bank branch nodes
- K: Set of transportation fleet
- T: Set of time periods
- P: Set of different banknotes

2.3.2.2 Parameters

- D_{it}: Demand of ATM i in period t in terms of monetary unit
- C_{ij}: Transportation costs between nodes i and j
- CO_2d: CO_2 emission costs per distance unit
- CO_2w: extra CO_2 emission per each 4000-banknote box
- fc: Fixed transportation costs
- CV_p: Per vehicle capacity
- h_p: Cost per unit banknote deposition p
- π_i: Cost per unit shortage per period for node i
- IO_{ip}: Initial p-type banknote inventory in branch i
- IC_p: Capacity of p-type banknote "in ATM"
- V_p: monetary value of the p-type banknote
- N: the number of branches.
- y_i: the optimal solution of the location model

2.3.2.3 Decision Variables

- I_{ipt}: Inventory level of the p-type banknote in branch i in period t
- S_{it}: Deficiency level for branch i in period t
- x_{ijkt}: A binary variable that equals 1 if vehicle k covers arc (i.j) in period t; it is 0, otherwise
- Y_{ikt}: A binary variable that equals 1 if branch i is visited in period t; it is 0, otherwise
- Q_{ijpkt}: Amount of p-type banknotes transported by vehicle k in period t in arc (i.j)
- Qd_{ipkt}: Amount of p-type banknotes delivered to branch i by vehicle k in period t

2.3.2.4 Modeling (Objective Functions and Constraints)

$$\min Z_1 = \sum_{i,k,t} fc\, x_{0ikt} + \sum_{i,j}\sum_{k,t} C_{ij} x_{ijkt} + \sum_{i,p,t} h_p I_{ipt}$$
$$+ \sum_{i,t} \pi_i S_{it} y_i \tag{2.16}$$

$$\min Z_2 = \sum_{i,j,k,t} x_{ijkt} \times CO_2 d \times d_{ij} + \sum_{i,p,k,t} Qd_{ipkt} \times CO_2 w \tag{2.17}$$

$$\sum_p I_{ipt} \times V_p - S_{it} = \sum_p I_{ip(t-1)} \times V_p + \sum_{p,k} Qd_{ipkt}$$
$$\times V_p - D_{it}, \quad \forall i \in J, t \in T, t > 1 \tag{2.18}$$

$$\sum_{j \in I} x_{ijkt} = \sum_{j \in I} x_{jikt} = Y_{ikt} \qquad \forall i \in J, t \in T, k \in K \tag{2.19}$$

$$\sum_{k \in K} Y_{ikt} \le 1 \qquad \forall i \in J, t \in T \tag{2.20}$$

$$\sum_{j \in I} Q_{jipkt} - Qd_{ipkt} = \sum_{j \in I} Q_{ijpkt} \qquad \forall t \in T, k \in K, p \in P \tag{2.21}$$

$$Q_{ijpkt} \le CV_p \times x_{ijkt} \qquad \forall i, j \in I, t \in T, k \in K, p \in P \tag{2.22}$$

$$\sum_{p,k} Qd_{ipkt} + I_{ip(t-1)} \le IC_p \qquad \forall i \in J, t \in T, t > 1 \tag{2.23}$$

$$Y_{0kt} \ge Y_{ikt} \qquad \forall i \in I, t \in T, k \in K \tag{2.24}$$

$$x_{iikt} = 0 \qquad \forall i \in I, t \in T, k \in K \tag{2.25}$$

$$US_{1kt} = 1 \qquad \forall t \in T, k \in K \tag{2.26}$$

$$US_{ikt} - US_{jkt} + 1 \le (N-1) \times (1 - X_{ijkt}) \qquad \forall t \in T, k \in K, (i, j) \in J, i \ne j \tag{2.27}$$

$$Y_{ikt}, x_{iikt} \in \{0, 1\}, \quad Q_{ijkpt}, Qd_{ikpt}, US_{ikt} \ge 0.integer, \forall i \in I, j \in J, t \in T, k \in K, p \in P \tag{2.28}$$

2.3.2.5   Modeling (Objective Functions and Constraints)

Equations (2.16) and (2.17) are the two objective functions: one minimizes the sum
of the fixed, transportation, money deposition, and deficiency costs and the other
minimizes the CO2 emissions. Equation (2.18) is the cash inventory balance at
ATMs, ensures that the ATM input and output is equal and Eq. (2.19) is a flow
conservation constraint for the vehicles. Constraint (2.20) ensures that a branch is
not visited by a vehicle more than once in a period. Equation (2.21) balances the
p-type banknote inventory for a vehicle that visits arc $(i.j)$ in period t. Constraint
(2.22) ensures that when vehicle k goes from node i to j in period t, it should
have a capacity greater than the banknotes it carries. Constraint (2.23) ensures
that each banknote inventory in period t is less than the ATM capacity, Constraint
(2.24) ensures the origin-departure trips, Eq. (2.25) determines the infeasible arcs,
and Eqs. (2.26) and (2.27) are the sub-tour elimination constraints, and finally,
Constraint (2.28) shows the types of variables.

2.3.3   ATM Location-Replenishment Integrated Model

This model combines the ATM location and replenishment models and the objective
function consists of four parts: (1) creating maximum coverage, (2) maximizing market share, (3) minimizing replenishment costs, and (4) minimizing CO_2
emissions. It is aimed to select the optimum branch locations considering the
replenishment costs, and since the selected optimum branch locations have not
been stablished yet, the model estimates the demand based on predictions. Its sets,
parameters, and decision variables are similar to those of the two previous models.
The variable N_{it} is a binary auxiliary variable to ensure that shortage and inventory
cannot be occurred concurrently, and M is an arbitrary large number.

2.3.3.1   Modeling

Objective functions and constraints are defined as follows:

$$max\, Z_1 = \sum_{j \in J} MS_j \times y_j \tag{2.29}$$

$$max\, Z_2 = \sum_{i \in I} \sum_{j \in J} a_{ij} \times y_j \tag{2.30}$$

$$min Z_3 = \sum_{i,k,t} fc \times x_{0ikt} + \sum_{i,j} \sum_{k,t} C_{ij} \times x_{ijkt}$$

$$+ \sum_{i,p,t} h_p \times I_{ipt} + \sum_{i,t} \pi_i \times S_{it} \times y_i \tag{2.31}$$

$$min Z_4 = \sum_{i,j,k,t} x_{ijkt} \times CO_2 d \times d_{ij} + \sum_{i,p,k,t} Qd_{ipkt} \times CO_2 w \times y_i \tag{2.32}$$

$$\text{All the constraints of the two previous models} \tag{2.33}$$

$$\sum_p I_{ipt} \leq M \times N_{it} \qquad \forall i \in J, t \in T \tag{2.34}$$

$$S_{it} \leq M \times (1 - N_{it}) \qquad \forall i \in J, t \in T \tag{2.35}$$

$$\sum_k Y_{ikt} \leq y_i \qquad \forall i \in J, t \in T \tag{2.36}$$

$$y_1 = 1 \tag{2.37}$$

$$x_{ij}, y_j, Y_{ikt}, X_{ijkt}, N_{it} \in \{0, 1\} \quad Q_{ijkt}, Qd_{ikt} \geq 0, integer \quad \forall i \in I, j \in J, k \in K, t \in T \tag{2.38}$$

Here, four constraints have been added to those of the two previous models; Constraints (2.34) and (2.35) show that when there is deficiency, inventory is zero and when there is inventory, deficiency is zero. Constraint (2.36) ensures that vehicles visit only those branches that have been selected from among the candidates and Eq. (2.37) ensures that the warehouse is always the vehicle starting point. Finally, Constraint (2.38) defines the variable types. It is noteworthy that the last term in the third objective function is a nonlinear term which can be linearized according to the linearization method explained in location model. It should not be left unmentioned that this term was not bilinear in the replenishment model since y_i was the input parameter in that model.

2.4 Numerical Experiments

2.4.1 Solution of the ATM Locating Model

This model is coded via OPL and CPLEX Script languages and solved using the CPLEX 12.8 Software. The data have been gathered from District 16 (Nazi Abad,

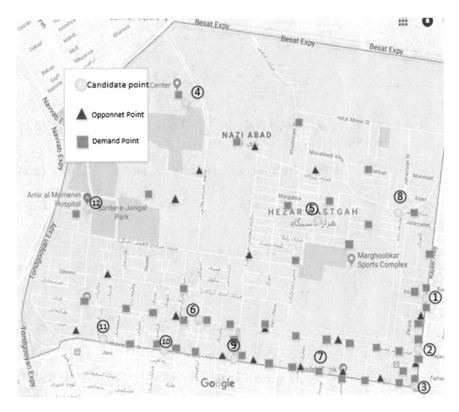

Fig. 2.1 Map of Nazi Abad neighborhood, Tehran

Tehran) that has all the amenities needed by a city and can be a good example
suitable for ATM studies. It hosts a cultural center, hospital, clinic, glass/crystal
factory, Tehran tobacco warehouse, park, and recreational centers. Since it has
two commercial streets that raise the district importance for banks, different bank
branches have been set up on them that show the high demand in the area. Figure 2.1
depicts the district map along with the demand points, candidate locations, and
branches of other banks.

Demand points (in blue) are 45 with different w_i values, branch-setup candidate
points (in yellow) are 12, and branches (own and rivals) (in red) are 17. First,
the x and y values for each demand/bank branch (own and rivals)/candidate
point are found using the Google Map and the related geographical latitude and
longitude coordinates, and then their distances from candidate points are calculated
orthogonally; for instance, the $(x_i, y_i) - (x_j, y_j)$ distance is found as follows:

$$d_{ij} = |x_i - x_j| + |y_i - y_j| \tag{2.39}$$

Table 2.3 Characteristics of candidate points

Characteristic	β_f
Access	0.4
Security	0.3
Physical factors/conditions	0.3

where S and T (taken from [18]) are 200 and 500 m, respectively. The candidate points' characteristics needed to calculate parameter A_j are shown in Table 2.3.

The utilization coefficient of each branch is taken to be 1 because the area texture is uniform. Values of α and λ (adjustment coefficients) are also 1 meaning that the demand point distance and branch attraction have equal effects.

To combine the two objective functions, use has been made of the lp-metric method which combines different-scale objective functions of a multi-objective problem (it has also been used to combine the objective functions of the replenishment and integrated models) [38]. In Eq. (2.40), Z_1max is the maximum value found for objective function Z_1 without considering objective function Z_2 and Z_2max is the maximum value found for objective function Z_2 without considering objective function Z_1.

$$lp - metric = [\omega_1 | \frac{Z_1max - Z_1}{Z_1max - Z1min}|^P + \omega_2 | \frac{Z_2max - Z_2}{Z_2max - Z2min}|^P]^{\frac{1}{P}} \quad (2.40)$$

where ω_1 and ω_2 are the objective functions' coefficients. Z_1max and Z_2max were calculated single objectively to be 1.926 and 48.57, respectively, and Z_1min and Z_2min were found to be 0. The extremum values have been calculated considering $p = 1$, and candidates 2, 6, 7, 10, and 12 have been selected as candidate points (Tables 2.4, 2.5).

Although candidate 8 has had market share compared to candidates 3, 6, and 7 it has not been selected because its coverage is less.

Table 2.4 Values of the market shares

Candidate (Size 12)	Value
1	0.155
2	0.162
3	0.136
4	0.377
5	0.267
6	0.152
7	0.148
8	0.195
9	0.144
10	0.427
11	0.150
12	0.651

Table 2.5 Values of the
objective functions

Lp-metric	Z1	Z2
0.389	1.54	39.39

2.4.2 Sensitivity Analysis of the Location Model

The first sensitivity analysis is done on objective functions' coefficients and w_1 and w_2 are multiplied by Z_1 and Z_2, respectively; sum of these two coefficients equals 1. Figure 2.2 shows two objective functions with increased w_1 and reduced w_2; as shown, an increase in w_1 increases the market share objective function and reduces the coverage objective function indicating that an increase in the market share does not necessarily increase the coverage and vice versa, which means that besides coverage, the increased market share also depends on such other factors as the type of demand existing in the region. For instance, although an ATM near a hospital covers only one demand point, it still provides more market share than one with more less-weighted demand points.

The second sensitivity analysis is done on α (attraction power from customer's point of view) and λ (power of customer distance from candidate point) in Eq. (2.10), meaning that the more is λ, the more is the importance of distance, and the more is α, the more is the importance of attraction.

Since the coverage objective function is not sensitive to α and λ and has similar values for different $\alpha - \lambda$ values, the lp-metric objective function too will vary only with the market share objective function. Figure 2.3 shows the graph of this function with increased α and reduced λ.

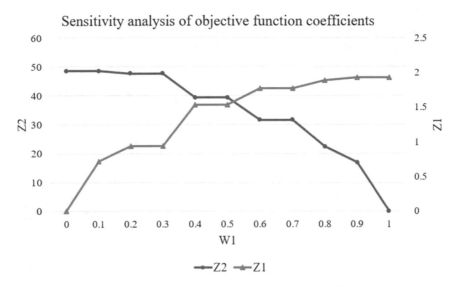

Fig. 2.2 Values of objective functions due to changes in their relative coefficients

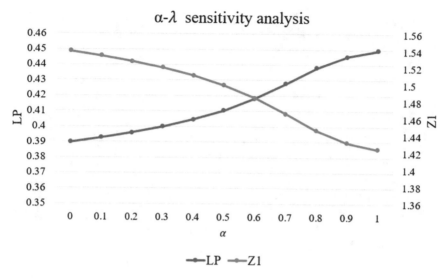

Fig. 2.3 Increased market share due to reduced α

Here, α is the power of the absolute attraction of candidate point (A_j^α) and λ is the power of the ATM distance from the demand point (a_{ij}^λ). As shown, with an increase in α, objective function Z_1 follows a descending trend showing that the distance of the candidate point from the demand point is more effective in the market share. This means that the smaller is the distance of the candidate point from the demand point, the greater will be the market share. For instance, an ATM in a main street but far from demand points will have less market share than one in an alley but close to demand points.

2.4.3   Solution of the ATM Replenishment Model

Proper refilling planning is possible after the optimum branch location is known. The daily interest is found by dividing the bank's annual interest (15%) by 365 (number of days in a year). Then, by multiplying this number by the ATM refilling period which is weekly, we can find the cost per unit money deposition.

$$\frac{0.15}{365} \times 7 = 0.0029$$

Costs per unit deficiency include those of the ATM cash non-availability, unsuccessful transactions, lost customers, lost commissions, and mal-advertisement for the bank (0.1 per unit). Transportation costs are in proportion to the distance covered; fuel consumption and depreciation costs are calculated per meter travelled and CO_2

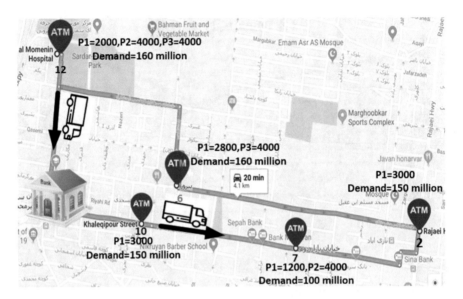

Fig. 2.4 Vehicle route in the first period

emission are due to distance covered and the freight (4000-banknote boxes) carried by the vehicle.

Being aware of the optimum ATM locations from the location model, we solve the replenishment model. Figures 2.4 and 2.5 show the replenishment model solution results (vehicle route and money injected into each branch) from the CPLEX Software in the first period (routes can be found using the values of X_{ijkt}). As seen in this figure the model utilizes only one vehicle which serves all stablished ATMs.

2.4.4 Solution of the Integrated Model

In order to show the value of integration, the integrated model was solved using the data from the two previous models; the lp-metric method was used to integrate the objective functions and candidate points 6, 9, 10, 11, and 12 were selected for the ATM setup. Except point 1 where the own bank is located, the rest of the points have been selected as the candidate points for ATM setups. Values of the four objective functions are also listed in Table 2.6.

Table 2.6 The values of the integrated model objective functions

Functions	Values
LP-METRIC	0.6176
Z1	1.4232
Z2	37.457
Z3	5,325,900
Z4	90.723

Table 2.7 Target function values with increasing carbon emission

$CO_2 d$	$Min Z4$	$Max Z1$	$Lp metric$
0.00025	965.04	0.70701	0.59568
0.00050	970.20	0.70701	0.59402
0.00060	976.74	0.70701	0.598
0.00065	978.41	0.70701	0.599
0.00070	1104.7	0.60249	0.597
0.00090	1107.9	0.60249	0.593

2.4.5 Sensitivity Analysis of the Integrated Model

Now, to check the carbon production effects on the profitability, we will first study the impacts of the increased carbon costs on the values of the coverage and market share functions (Table 2.7).

An increase in the carbon emission increases the value of the lp-metric objective function until the ATM optimal location is changed which will reduce the values of the coverage and market share objective functions and the lp-metric function. Another increase in the carbon emission increases the value of the lp-metric function; hence, those candidate points that are closer to the central warehouse will be selected for the ATM setup.

2.4.6 Comparison of the Hierarchical Approach and the Integrated Model

Hierarchical approach in this paper is defined as follows: The location model is separately solved to select the best ATMs locations. Then, the selected ATMs are stablished and the real demand of the ATMs is measured after a while and based on the real withdrawals. After that, being aware of the ATMs demand, the replenishment model is configured and solved to obtain the optimal routing and cash deliveries accordingly. While in the integrated approach, the decisions of the ATMs locations, and the vehicle routing and deliveries are all determined concurrently. Therefore, the real values of the demand are not reviled in the integrated model contrary to the hierarchical approach. So, in the integrated approach one should rely on the predicted demand instead of the real demand. For this comparison, we first find the values of the objective functions by a hierarchical approach, then use the

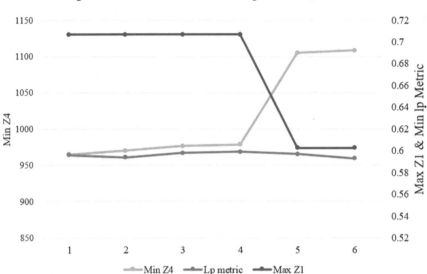

Fig. 2.5 Values of the objective functions due to increased carbon emission

same data to solve the integrated model, and finally compare the related results. Here, since the candidate points demand, in the integrated model, affects proper selection of ATM locations, it was assumed to be equal for all candidates and both methods were solved by this assumption. In the hierarchical approach, we first found the optimum ATM locations and then solved the replenishment model with equal demand for the selected points.

$$Z_1 = 1.54 \qquad Z_2 = 39.39$$
$$Z_3 = 7{,}272{,}600 \qquad Z_4 = 140.58$$

Values of the three objective functions are summed up by the lp-metric method.

$$lp - metric = [\omega_1 |\frac{Z_{(1max)} - Z_1}{Z_{(1max)} - Z_{(1min)}}|^P + \omega_2 |\frac{Z_{(2max)} - Z_2}{Z_{(2max)} - Z_{(2min)}}|^P$$
$$+ \omega_3 |\frac{Z_3 - Z_{(3min)}}{Z_{(3max)} - Z_{(3min)}}|^P + \omega_4 |\frac{Z_4 - Z_{(4min)}}{Z_{(4max)} - Z_{(4min)}}|^P]^{\frac{1}{P}}$$
$$= \frac{1.9258 - 1.54}{1.9258} + \frac{48.5729 - 39.39}{48.5729} + \frac{7{,}272{,}600 - 5{,}258{,}400}{5{,}258{,}400}$$
$$+ \frac{140.58 - 89.373}{89.373} = 1.3454$$

Table 2.8 The values of the objective functions of both methods

		Market share (Z_1)	Coverage (Z_2)	Refilling (Z_3)	Cost $Co_2(Z_4)$	Total (*Lp-metric*)
Hierarchical approach	Objective functions	1.54	39.39	7,272,600	140.58	1.3454
	Selected points	2	6	7	10	12
Integrated model approach	Objective functions	1.4232	37.457	5,325,900	90.723	0.6176
	Selected points	6	9	10	11	12

In the hierarchical approach, values of the coverage and market share objective functions as well as the replenishment costs are greater compared to the integrated model. It means that in the integrated model, the ATM replenishment plans have been concurrently investigated besides maximizing the market share and coverage causing a decrease in the overall minimization objective function compared to the hierarchical approach. This observation demonstrates the added-value of the integration that is proposed by the location-replenishment integrated model. It should be noted that, this observation is true when the demand in integrated model has been estimated accurately. When the demand estimation is erroneous, results of the hierarchical approach may be better. To find the acceptable percent error to use the integrated model, we first solve it with the predicted demand, and since demand is only a prediction, the value found for the replenishment objective function will not be real. To find the real value, the replenishment model is solved using the points found from the integrated model and the exact demand which is somewhat different from the predicted one. The result is then used in the integrated model to find the value of the lp objective function. Hence, it is possible to say how better (with what percent error) the solution of the integrated model is compared to the hierarchical approach (Table 2.8).

According to Table 2.9, up to 15% error, the integrated model performs better than the hierarchical approach. Hence, it can be concluded that if the demand

Table 2.9 Estimated percentage of acceptable error

Percentage error	Lp-Integrated model	Lp-Hierarchical approach
2%	0.7295	1.3454
5%	0.9866	1.3454
10%	1.193	1.3454
15%	1.3271	1.3454
20%	1.5041	1.3454

considered for the integrated model is less erroneous than 15%, the integrated model will perform better than the hierarchical approach.

2.5   Conclusions

In this study, three models: (1) ATM location, (2) replenishment planning, and (3) an integration of the two were presented. The first one addressed the partial coverage and market share considering the rivals and attraction of the candidate points, the second one considered different-value cash injection into ATMs plus the pollution routing problem, where the costs of transportation, deficiency, storage, deposition, and carbon emissions are minimized, and the third one determined the optimum ATM locations so as to minimize the replenishment and carbon emissions in addition to maximizing the coverage and market share concurrently. In locating ATMs, the coverage and attraction were, for the first time, combined in a competitive environment, in replenishment side, it was also for the first time that cash injection was planned for different-type banknotes considering carbon emissions, and, finally, the two were combined to yield an integrated model. The results showed that the coverage and market share are not necessarily directly related; in other words, an increase in the market share does not necessarily increase the coverage or an increase in the coverage does not necessarily increase the market share which means that the market share is more affected by the type of the demand to be met. Again, the ATM proximity to demand points creates more market share than such attractions as being located in a main street. We also found that an increase in carbon emissions (i.e. vehicle engine technology, e.g. Euro 4, 5, etc.) will reduce the coverage and market share; hence, points are selected for ATMs that are closer and have lower CO_2 emission in the replenishment process which will respect the environmental concerns of the people and governments. The comparison of the hierarchical and integrated approaches also indicates that if the estimation of the integrated model demand is correct and less erroneous, it can provide better solutions than the hierarchical approach. Results showed that this error was nearly 15% meaning that if the demand prediction error is less than 15%, the integrated model will yield better results than the hierarchical approach and have a meaningful positive impact on the economic and environmental aspects, concurrently. This research can be improved if the following are considered for future studies; (1) solving the integrated model along with the demand prediction modeling, (2) considering demand uncertainty and/or elasticity in the replenishment model, (3) identifying low-market share, low-coverage ATMs for closure, besides locating optimum ones, (4) considering competition dynamically, and finally (5) enabling the model to stablish the mobile ATMs.

References

1. El-Haddad, A.B., Almahmeed, M.A.: ATM banking behaviour in Kuwait: a consumer survey. Int. J. Bank Marketing **10**(3), 25–32 (1992)
2. Marshall, J.J., Heslop, L.A.: Technology acceptance in Canadian retail banking: a study of consumer motivations and use of ATMs. Int. J. Bank Marketing **6**(4), 31–41 (1988)
3. Swinyard, W.R., Ger Ghee, L.: Adoption patterns of new banking technology in Southeast Asia. Int. J. Bank Marketing **5**(4), 35–48 (1987)
4. Trautman, W.B.: A framework for regulating automated teller machine technology. J. Policy Anal. Manag. **12**(2), 344–358 (1993)
5. Ranković, M., Vasković, V.: The economic models for the ATM network implementation. IPSI BgD Trans. Adv. Res **5**(2), 16–21 (2009)
6. Iberahim, H., Taufik, N.M., Adzmir, A.M., Saharuddin, H.: Customer satisfaction on reliability and responsiveness of self service technology for retail banking services. Procedia Econ. Finance **37**, 13–20 (2016)
7. Almossawi, M.: Bank selection criteria employed by college students in Bahrain: an empirical analysis. Int. J. Bank Marketing **19**(3), 115–125 (2001)
8. Miliotis, P., Dimopoulou, M., Giannikos, I.: A hierarchical location model for locating bank branches in a competitive environment. Int. Trans. Oper. Res. **9**(5), 549–565 (2002)
9. Aldajani, M.A., Alfares, H.K.: Location of banking automatic teller machines based on convolution. Comput. Ind. Eng. **57**(4), 1194–1201 (2009)
10. Li, Y., Sun, H., Zhang, C., Li, G.: Sites selection of ATMs based on particle swarm optimization. In: 2009 International Conference on Information Technology and Computer Science, vol. 2, pp. 526–530. IEEE, Piscataway (2009, July)
11. Alhaffa, A., Al Jadaan, O., Abdulal, W., Jabas, A.: Rank based genetic algorithm for solving the banking ATM's location problem using convolution. In: 2011 IEEE Symposium on Computers and Informatics, pp. 6–11. IEEE, Piscataway (2011, March)
12. Mimis, A.: A geographical information system approach for evaluating the optimum location of point-like facilities in a hierarchical network. Geo-spatial Inf. Sci. **15**(1), 37–42 (2012)
13. Mourad, M., Galal, N.M., El Sayed, A.E.: Optimal location-allocation of automatic teller machines. In Conference: World Academy of Science, Engineering and Technology, UAE, Dubai, vol. 61 (2012)
14. Tabar, M.M., Bushehrian, O., Moghadam, R.A.: Locating ATMs in urban areas. Int. J. Comput. Sci. Eng. **5**(8), 753 (2013)
15. Ehsani, A., Danaei, A., Hemmati, M.: A mathematical model for facility location in banking industry. Manag. Sci. Lett. **4**(9), 2097–2100 (2014)
16. AwaghadeII, S., DandekarI, P., RanadeII, P.: Site selection and closest facility analysis for automated teller machine (ATM) centers: Case study for Aundh (Pune), India. Int. J. Adv. Remote Sens. GIS Geogr. **2**(1), 19–29 (2014)
17. Bilginol, K., Denli, H.H., Şeker, D.Z.: Ordinary least squares regression method approach for site selection of automated teller machines (ATMs). Procedia Environ. Sci. **26**, 66–69 (2015)
18. Trang, P.T., Sonb, N.L.N., Giangc, P.T.: The Influence of ATM location characteristics on ATM usage in Vietnam. Int. J. Adv. Eng. Manag. Res. **4**(03), 2019 (2019)
19. Kumar, P., Walia, E.: Cash forecasting: an application of artificial neural networks in finance. IJCSA **3**(1), 61–77 (2006)
20. Bilir, C., Doseyen, A.: Optimization of ATM and branch cash operations using an integrated cash requirement forecasting and cash optimization model. Bus. Manag. Stud. Int. J. **6**(1), 237–255 (2018)
21. Kurdel, P., Sebestyénová, J.: Routing optimization for ATM cash replenishment. Int. J. Comput. **7**(4), 135–44 (2013)

22. Simutis, R., Dilijonas, D., Bastina, L.: Cash demand forecasting for ATM using neural networks and support vector regression algorithms. In: 20th International Conference, EURO Mini Conference, "Continuous Optimization and Knowledge-Based Technologies" (EurOPT-2008), Selected Papers, Vilnius, pp. 416–421 (2008, May)

23. Brentnall, A.R., Crowder, M.J., Hand, D.J.: A statistical model for the temporal pattern of individual automated teller machine withdrawals. J. R. Stat. Soc. C (Appl. Stat.) **57**(1), 43–59 (2008)

24. Teddy, S.D., Ng, S.K.: Forecasting ATM cash demands using a local learning model of cerebellar associative memory network. Int. J. Forecasting **27**(3), 760–776 (2011)

25. Armenise, R., Birtolo, C., Sangianantoni, E., Troiano, L.: Optimizing ATM cash management by genetic algorithms. Int. J. Comput. Inf. Syst. Ind. Manag. Appl. **4**, 598–608 (2012)

26. Baker, T., Jayaraman, V., Ashley, N.: A data-driven inventory control policy for cash logistics operations: An exploratory case study application at a financial institution. Decision Sci. **44**(1), 205–226 (2013)

27. Darwish, S.M.: A methodology to improve cash demand forecasting for ATM network. Int. J. Comput. Electr. Eng. **5**(4), 405 (2013)

28. Venkatesh, K., Ravi, V., Prinzie, A., Van den Poel, D.: Cash demand forecasting in ATMs by clustering and neural networks. Eur. J. Oper. Res. **232**(2), 383–392 (2014)

29. Zandevakili, M., Javanmard, M.: Using fuzzy logic (type II) in the intelligent ATMs' cash management. Int. Res. J. Appl. Basic Sci. **8**(10), 1516–1519 (2014)

30. Ekinci, Y., Lu, J.C., Duman, E.: Optimization of ATM cash replenishment with group-demand forecasts. Expert Syst. Appl. **42**(7), 3480–3490 (2015)

31. Tsyganov, A.: Cash withdrawal from ATMS as long memory time series. Procedia Comput. Sci. **88**, 459–462 (2016)

32. Van Anholt, R.G., Coelho, L.C., Laporte, G., Vis, I.F.: An inventory-routing problem with pickups and deliveries arising in the replenishment of automated teller machines. Transp. Sci. **50**(3), 1077–1091 (2016)

33. Larrain, H., Coelho, L.C., Cataldo, A.: A variable MIP neighborhood descent algorithm for managing inventory and distribution of cash in automated teller machines. Comput. Oper. Res. **85**, 22–31 (2017)

34. Zhang, Y., Kulkarni, V.: Automated teller machine replenishment policies with submodular costs. Manuf. Serv. Oper. Manag. **20**(3), 517–530 (2018)

35. Mohammadi, S., Mirzapour Al-e-Hashem, S. M. J., Rekik, Y.: An integrated production scheduling and delivery route planning with multi-purpose machines: A case study from a furniture manufacturing company. Int. J. Prod. Econ. **219**, 347–359 (2020)

36. Nakanishi, M., Cooper, L.G.: Parameter estimation for a multiplicative competitive interaction model—least squares approach. J. Marketing Res. **11**(3), 303–311 (1974)

37. Timajchi, A., Mirzapour Al-e-hashem, S. M. J., Rekik, Y.: Inventory routing problem for hazardous and deteriorating items in the presence of accident risk with transshipment option. Int. J. Prod. Econ. **209**, 302–315 (2019)

38. Mirzapour Al-e-hashem, S.M.J., Rekik, Y., Hoseinhajlou, E.M.: A hybrid L-shaped method to solve a bi-objective stochastic transshipment-enabled inventory routing problem. Int. J. Prod. Econ. **209**, 381–398 (2019)

Chapter 3
Modelling a Future Routing Concept for the Urban Air Mobility

Moritz Hildemann and Carlos Delgado

Abstract New advances in electrical vertical take-off and landing (eVTOL) air-crafts require adaptations in the management of the future urban air space. The new mode of green air transportation offers opportunities like fast commuting times and more flexibility. However, it involves drawbacks, for example an increased noise pollution or safety violations. These conflicts need to be effectively managed. We propose a solution that satisfies citizens, relieves urban transport systems and reduces environmental impacts in urban areas. In a first step, information from the flight laws are used to model the restricted air space of the future. These air space restrictions are combined with the proposed protected areas designated for preserving the life quality of citizens. In a next step, a route network is proposed taking the following objectives into account: minimizing the networks flight path length, minimizing the noise pollution for the citizens and minimizing risk in emergency cases. Lastly, we improve the restricted air space and network to be adaptable and scalable. The model is designed to find optimal solutions for different types of urban areas and is adjustable to temporal changes within each urban area. These temporal conditions arise from changing safety requirements, a changing consumer demand and a changing noise level. We introduce the site planning for vertihubs and vertiports for landing an eVTOL aircraft in this temporally changing environment. Using New York City as the study area, our research demonstrates how the flight network for Urban Air Mobility can be designed to meet safety requirements while increasing the citizens' willingness to engage with the new green mode of transportation.

M. Hildemann (✉)
Institute of Geoinformatics, Westfälische Wilhelms-Universität, Münster, Germany
e-mail: jhildema@uni-muenster.de

C. Delgado
Universidade Nova de Lisboa, Lisbon, Portugal
e-mail: m20181032@novaims.unl.pt

© Springer Nature Switzerland AG 2020
H. Derbel et al. (eds.), *Modeling and Optimization in Green Logistics*,
https://doi.org/10.1007/978-3-030-45308-4_3

3.1   Introduction

Several start-up and grown-up companies like Uber [1], Lilium Aviation [2], Volocopter [3] and Airbus [4] among others are planning to transform the future urban air space by introducing a new sustainable mode of transportation: electrical vertical take-off and landing aircrafts. These aircrafts are environmentally friendly as they are electrically driven and engineered to be quieter than helicopters that run on fuel [5]. The additional green mode of transportation can relieve the road traffic, especially in Mega cities. And this can avoid pollution in terms of dust exhaust or excessive noise. Moreover, the respective transportation time can be reduced significantly by using this new technology [6].

To realize this futuristic idea of flying cars in urban areas, many requirements need to be fulfilled. One of them is the guarantee of safety [6], and another is the consent of the society [7, 8]. This chapter discusses how these requirements can be considered when the flight paths are modelled. Concepts of the division and separation of the future urban air space exist [8], as well as extensive theory about flight network structures [9]. This chapter models the suggested air space division and computes the flight network in different cities structures with different flight laws. This work gathers the different requirements in the study area of New York Manhattan to point out the challenges and the key issues in order to support the launch of this new mode of green transportation.

The first part is dedicated to the question how to model the restricted air space by taking into account the air space restrictions of the local flight law [10, 11] as well as to model the protected air space. This protected air space is included in the urban planning to increase the probability of getting the acceptance of the society. This includes the avoidance of flight routes above points and areas of interest to the citizens. We will discuss which areas might be considered as a protected air space and also how this can be modelled in different cities.

The second part's objective is to optimize flight paths of the non-restricted and non-protected air space. We explain and illustrate how the optimal routes for different criteria can be computed. These criteria can be the minimization of the flight path length, the minimization of the noise disturbance and the minimization of danger in emergency situations, which are examples for different optimization criteria for the flight network of the electrical aircrafts. The third part contributes to the question: How can we optimize the network structure for the given air space restrictions and optimization criteria? This includes a discussion about a possible network structure that can be modelled to be dynamic and expendable [9]. Furthermore, the site planning of the starting and landing points on rooftops and its relevance to the network structure are elaborated.

Lastly, the question is approached, how all the discussed methods can be combined in one model to produce the optimal flight network with different inputs. The given parameters can contain information about the air space restrictions, air space protections and the optimization criteria. These parameters can be specified by experts like the city representatives to compute the best flight network for any specific urban area.

3.2 Modelling the Restricted Air Space in Urban Areas

Flight laws regulate urban areas now to ensure safe and risk-free approaches and departures of planes to the airports. Small aircrafts like the planned air taxis as well as unmanned systems (UAS) face these restrictions[10, 12]. National and international aerial space-regulating institutions like the DLR [8], CAA [12] or the NASA [6] developed concepts to integrate these aircrafts [8]. The concepts propose how the airspace might be divided and limited in the three-dimensional air space, to ensure the safety of the citizens. These may be used as a consultation for adaptations of the flight laws in the future [13]. Information of restricted areas can be modelled with three-dimensional geofences [8]. Restricted areas by the actual flight laws [14] in combination with the proposed protected areas, for example embassies or national institutions, can be combined and modelled as geofences with different heights depending on the restricted or protected area.

3.2.1 Legal and Soft Restrictions for the Flight Paths

The information for the flight restrictions derive from aerial space agencies. In this chapter, we present the example of Manhattan in New York, and therefore, the model complies with the flight restrictions of the Federal Aviation Administration (FAA) [14] from the United States of America. Furthermore, the restrictions differ for different types of aircrafts: unmanned aircraft systems, ultra-light vehicles, aircrafts with and without communication possibilities, fixed-wing aircrafts and helicopters, including weighted shift-control vehicles [14]. The regulations for the sample area and the example vehicle (VTOL aircrafts) are the following:

Sec. 91. 119 FAA, Paragraphs c, d

(c) Over other than congested areas: An altitude of 500 feet above the surface, except over open water or sparsely populated areas. In those cases, the aircraft may not be operated closer than 500 feet to any person, vessel, vehicle or structure.

(continued)

(d) Helicopters, powered parachutes and weight-shift-control aircraft. If the
     operation is conducted without hazard to persons or property on the
     surface:

     [...]
     (2) A powered parachute or weight-shift-control aircraft may be operated at less
     than the minimums prescribed in paragraph (c) of this section.

These legal restrictions can be modelled as static geofences, for example
with the support of Geographical Information Systems (GIS). In the study area
Manhattan already exists a helicopter path, which complies with 119 FAA a(1)A.
The helicopter routes are planned above open water regions to guarantee safety and
to minimize annoyance [14]. The electrical aircrafts are not as noisy and large as
a traditional helicopter, so we propose additional routes for this kind of aircrafts
that are not planned only above water bodies. To maximize the probability of high
acceptance of the affected population, we propose to model protected areas as soft
restrictions in addition to the legal/hard restrictions. We propose to consider schools,
universities, graveyards and recreational parks as protected areas. We want to point
out that these protected areas can be adapted by the responsible decision-makers.
It also guarantees the adaptability of the model to different types of urban areas.
For example, in Manhattan, recreational parks are rare and need to be protected in
our point of view. Whereas in urban areas that have many recreational parks, this
landuse type does not need to be generally protected.

3.2.2 Generating 3D Geofences to Model the Restricted Air Space

The concept of the German Space Agency DLR suggests a combination of static
and temporal/dynamic geofences for the future urban air traffic management [8].
We model the static geofences with the mentioned restrictions over graveyards,
airports, recreational parks or universities [12, 15] with the tool ArcGIS Pro from
Esri [16]. The practical approach is explained in detail in a conference paper [17] for
Geoinformation Science. In this chapter, we go into more detail about the conceptual
idea with the results of modelling a 3D network for the new mode of transportation.
We therefore explain in this work the necessary technical background and workflow
to grasp the idea of the approach. The first step is to model the ground surface with
all buildings and their height[18] and to add the minimal height restrictions above
the populated area of 500 feet[10]. On top of this, the restricted geographic areas
such as educational centres, hospitals, embassies and cemeteries are subsequently
used to calculate 3D geofences. Due to the fact that large landuse areas are treated
different than buildings, the geofences for the landuse areas parks, graveyards

Table 3.1 Vertical and horizontal restrictions in metres

Landuse	Vertical restriction	Horizontal restriction
Airport	600	Dynamic
Embassies	300	300
Hospitals	300	300
Universities	200	300
Parks	300	100
Graveyards	300	100
Recreational areas	300	100
Rooftops	152.4	–

and recreational areas differ in the height. The different height restrictions are listed in Table 3.1. The highest restriction is the 25,000 feet distance line to the airports declared by the Federal Aviation Administration [10]. These are especially protected, and therefore, these geofences are modelled with the height of 600 m. As exception, we also modelled a special way in the geofences as a small path to and from the airports to enable a connection to the airport. This might be seen as a highly supervised channel and they are only proposed by us, but this exception is actually not existing. The data for modelling the geofences is Open Source:

- Open Street Map Data [19],
- Flight obstacle Maps [10],
- Rooftop heights from Open Data NY [18].

The model is available on GitHub in a toolbox format from the software Esri. If licenses are available and the necessary data downloaded, every reader is welcome to reproduce the results or to run the model on a different study area.[1]

3.2.3   Generate the Minimal Flight Height Plane

We create a minimal flight height plane with a minimum of 152.4 m and a the maximum flight height of 213.36 m. To save as much energy as possible for the aircrafts, the minimal flight height is the desired height for operating. The minimal flight height plane is the combination of the non-restricted air space above the minimal flight height of 152.4 m above the surface and the restricted air space with the 3D geofences as barriers. The format of the 3D geofences are extruded shapes (vector format), and the surface model is a grid (raster format)[20]. This grid is a surface where each cell covers a square of 100 times 100 m on the digital earth surface model and includes a height value of the covered area. In this particular

---

[1]The 3D geofences are reproducible with Model_1_3D_Geofences in the toolbox available on https://github.com/mohildemann/Urban-Air-Mobility-Routing.

case, the height of the buildings that are built on top of the surface is added to this value. At this point, we have the minimal height plane without the geofences of the restricted areas.

To create the minimal flight height plane including the information of the restricted air space, the shapes of the 3D geofences also need to be conversed to raster format with the same cell size. If a geofence exists at the same raster cell, it replaces the surface height value with the stated heights in Table 3.1. The conversion means that the information of the circle shapes are assigned to the underlying raster cell. This leads to a loss of information, as the shape is being pixelized, comparable to making a picture where each pixel is 100 times 100 m. To do this conversion, we used the interpolation method inversed distance weight (IDW). This estimates the cell value by taking into account all neighbouring cells within a predefined distance and giving closer cells a higher weight to estimate the cell value than the more distant cells [20].

The generated IDW surface, illustrated in Fig. 3.3,[2] visualizes the combined information of the restricted and non-restricted areas. The open water regions with an assumed flight height of 100 m are visualized in red, followed by the basic heights of the ground surface height plus the height of the building plus 152.4 m, which are in orange. The yellow parts represent the 3D buffers with a lower vertical restriction (200 m) than the blue parts (300–600 m) of the map, which represent the no-fly areas. The yellow parts can be overflown, but the aircraft would need to ascend to the minimal flight height of 200 m. As the maximal flight height is 213.36 m (700 feet), the blue parts of the map are completely restricted. This guarantees that the soft and hard restrictions described in Sect. 3.2.2 are modelled with this methodology. The generated surface can now be used as a cost surface. This means that the value of each raster cell represents a cost value.

3.3 Model Best Routes for Different Criteria in the Non-restricted Air Space

Now, the created least cost surface in Sect. 3.2.2 is used as input to an algorithm, which finds the path with the lowest accumulated cost by following the cost surface. In the scenario of urban air taxis, the aircrafts will try to avoid all geofences with soft restrictions while completely avoiding the hard restrictions. The best route is the smallest accumulated sum of the costs between the starting and landing points, where the cost is the distance of each raster cell to their neighbour cells in the IDW. That is calculated using the Cost Connectivity Tool from ArcGIS Pro.[3] When the

---

[2]The IDW is reproducible with Model_2_Least_Cost_Surface in the toolbox available on https://github.com/mohildemann/Urban-Air-Mobility-Routing.

[3]Cost Connectivity Tool bases up on the computation of the Minimal Spanning Tree Algorithm [21].

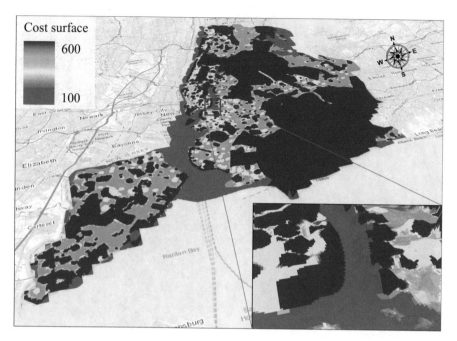

Fig. 3.1 Least cost surface (IDW)

starting and landing points, the so-called hubs, are defined, the cost connectivity tool calculates the shortest cost paths from each hub to the other ones. All least cost paths combined are called the least cost network. The following sub-chapters explain how the least cost network is calculated with different underlying costs, depending on the chosen criteria. The criteria can be only the shortest distance, or a combination of the shortest distance with other criteria, for example to minimize the noise pollution or to minimize the danger in emergency situations.

3.3.1   Minimize Total Distance

To illustrate the least cost paths, we can imagine a person who is hiking in the mountains. To get from a starting point to the desired destination, the hiker can choose if he surrounds some mountains as obstacles. The probability of dodging the mountains increases, if the mountain is very high and steep. However, the hiker thinks twice, if the detour is too long, so he might consider the risk of climbing the mountain less than arriving in the dark. Our model includes a parameter of the cost factor that multiplies the cost value of the artificial surface (IDW). In the example of the hiker, we exaggerate the height and the steepness of the mountains as obstacles. This allows to include a parameter in our model that increases the cost of overflying

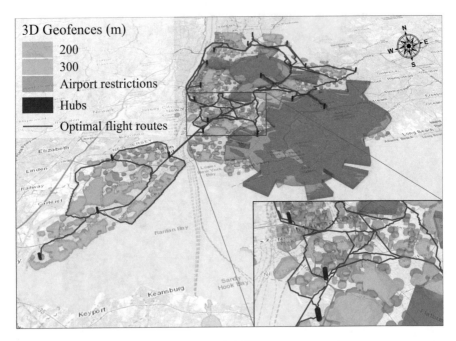

Fig. 3.2 Least cost network and 3D geofences in NYC

geofences. This is important for the geofences that model the protected areas that are lower than the maximal flight height, like the hospitals with a height of 200 m.

That parameter can be set to a low value to minimize flight times and energy consumption, but a higher probability of overflying protected areas if the detour is too long. If that parameter is set to a higher value, the probability of dodging the geofences and also the path length increase. The parameter allows the optimal and highly accurate route calculation for different types of urban areas. The model used to create the network illustrated in Fig. 3.2 had a factor of 30.[4] This was considered as the optimal parameter for the city of New York and enforces a very high probability of dodging the geofences rather than flying above. The hubs were selected manually, and open areas close to railway, metro and bus stations were preferred to ensure a good connectivity to the public transport.

[4]The least cost network reproducible with Model_3_Least_Cost_Network in the toolbox available on https://github.com/mohildemann/Urban-Air-Mobility-Routing.

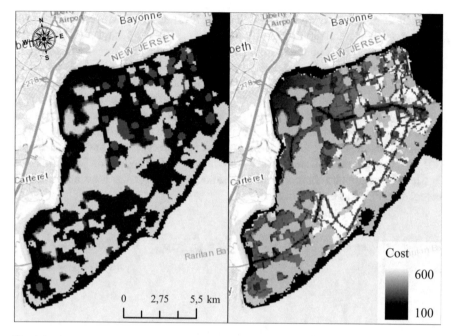

Fig. 3.3 Comparison of least cost surfaces for distance and noise

3.3.2 *Minimize Noise Pollution*

We considered another criteria than distance as crucial as a common aim of the responsible aviation agencies and health organizations is to reduce the negative impact of noise on the citizens [22–24]. One of the most important advantages of electrical vertical take-off and landing (VTOL) is the reduced noisiness compared to fuel-driven helicopters. Uber Elevate [1] expects the vehicles to create a quarter of the noise of the quietest helicopter on the market. Moreover, the Engineering Director of Aviation Marc Moore of Uber ascribes the vehicles to have a sound with a higher pitch, which expects the sound to be "blending into the hum of car traffic in cities rather than rumbling on over a longer distance and rattling windows" [5]. But the technical requirements by Uber [1], as one example, require the aircrafts to be less noisy than 62 dBA, but these requirements are not be met yet [25].

Therefore, the aim of this work is to introduce a routing, which takes the existing noise of the road traffic and air traffic into account. The flight paths are planned to be in places with already existing high traffic noise, rather than producing noise in quite areas. This avoidance of noise in calm areas, for example residential areas, is desired by the organizations that deal with the effect of noise on citizens [24, 26]. The planning of flight paths in noisy areas shall lead to a minimal increase in the annoyance of the affected citizens. To achieve the aim, we include noise data of the study area in the existing least height plane created in Sect. 3.3.1. The minimal height plane is manipulated to create a different cost surface for the least cost network calculation. To create a cost surface that leads to a network with a smaller negative effect on the citizens, the cost will be higher increased in locations where the noise is less than in noisy places. The second objective is to introduce a parameter in the calculation, which allows the factor noise to be given higher or less importance for the computation of the least cost network. If a high value is chosen by the possible end user, the importance of the factor noise increases and the importance of the criteria least distance therefore decreases. The data for a weighted, average sound level for the day from aviation and interstate road noise in the year 2014 is available at the United States Department of Transportation [27]. This data is used to manipulate the cost surface, which is used for the least cost network calculation that was created in Sect. 3.2.3.

$$
C_{new} = \begin{cases} C_{old} + \left((C_{max} - C_{old}) * W * \left(1 - \frac{N_{px} - N_{min}}{N_{max} - N_{min}} \right) \right) & \text{if } Pix_{Land} \neq water \\ 100 & \text{if } Pix_{Land} = water \end{cases}
$$

(3.1)

$$
C_{new} = \begin{cases} C_{max} & \text{if } C_{new} > C_{max} \\ C_{new} & \text{else} \end{cases}
$$

(3.2)

where

C_{new}:   new pixel value in the cost surface;
C_{old}:   old pixel value in the cost surface;
C_{max}:   maximum flight height (213.36 m);
W: weight (impact factor noise);
N: noise value at that raster cell;
N_{min}:   minimum noise value in the study area;
N_{max}:   maximum noise value in the study area and
Pix_{Land}:   landuse of the raster cell.

The recalculation of the cost surface includes the noise by increasing the cost depending on the noise level of the location. The lower the noise level, the higher is the added cost to the cost surface as expressed by Eq. (3.1). If the current position is above a water body, the cost will be recalculated to the minimum cost of 100 m as expressed by Eq. (3.2). The maximum cost is the maximum flight height of 213.36 metres. In this way, new routes can be created, which tend to follow water bodies and high levels of noise. However, in order to achieve this objective, we considered an impact factor that can be adjusted by the decision-maker.[5]

Table 3.2 Comparative table of the average noise and total length for each of the networks obtained

Type of least cost network	Average noise (dBA)	Total length (km)
Without considering the noise map	49.39	622
With an impact factor of 1	50.64	629
With an impact factor of 2	51.14	684

An overview of each of the calculated networks is observed in Fig. 3.4. All paths try to avoid the 3D geofences, and the trajectories vary with the chosen impact factor. This is mainly due to the change in the least cost surface, which is displayed in Fig. 3.3. It shows that the least cost surface, which takes the noise factor into account (right), is more complex than the previous cost surface with only the minimal flight height (left). The cost surface including the noise is not plane in places without geofences as the surface without the noise cost. Looking more closely at the behaviour of the new networks, it is confirmed that the algorithm is sensitive to an increase in the noise impact factor. This is reflected in Figs. 3.6 and 3.7, where it can be observed that the path of the route tries to follow the streets and highways, which have a higher level of noise. The difference per factor is displayed in Table 3.2.

Therefore, this comparative table reflects a series of alternatives for the design of networks for air taxis. In any case, it is necessary that each of these given alternatives is studied in greater depth by experts or decision-makers. It is important to mention that within the developed networks the noise levels do not exceed those maximum allowed values, according to the World Health Organization [24, 28], Civil Aviation Authority [22] and Office of noise abatement and control [26]. These organizations recommend that noise levels during the daytime and day–night do not exceed 55 dB in order to prevent interference and annoyance.

3.3.3 Minimize Danger in Emergency Situations

If the safe flight to a port or an active hub is infeasible, an emergency landing site has to be approached. Several options exist to gain crucial information regarding hazardous situations [29]. One option for vertical take-off and landing aircrafts is to identify a landing site using on-board sensors. Visual sensors are used to collect data

---

[5]The least cost network for minimizing noise pollution is reproducible with Model_4_Noise_Assessment in the toolbox available on https://github.com/mohildemann/Urban-Air-Mobility-Routing.

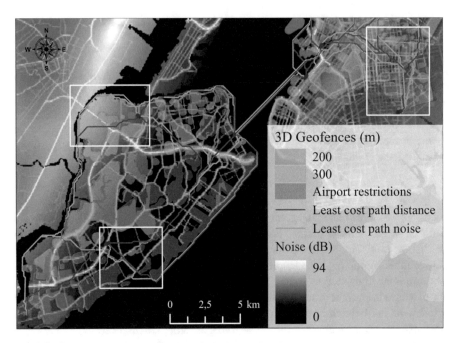

Fig. 3.4 Network comparison of least costs distance and noise

with cameras or Light Detection and Ranging (LiDAR), which are processed in near real time with object recognition algorithms [30]. These could identify obstacles or also human beings on the ground, and then these obstacles can be avoided. The disadvantages of this option are that the safest landing site can be behind the aircraft or not visual to the sensor, because obstacles hide possible landing sites [29]. Another option is to plan the landing sites in advance for each possible location in the network [31]. The sites can be planned and the information stored. Hence, we included emergency landing sites in our model, in which the longest distance to an emergency landing site is 3000 m from each location in the study area. Emergency situations are prioritized over the discomfort of the citizens; therefore, the emergency landing sites can be located within restricted areas. Consequently, emergency paths can also cross the geofences.

We replace the existing cost surface in emergency situations with the current population density at the location for a specific time range. The cost surface including the population density can therefore change rapidly over the time of the day as well as per day of the week. An interesting visualization of the hourly estimated population per weekday for Manhattan can be seen in an interactive map from Fung [32]. The used data for the estimation of the population was data gathered from the public transportation. We used the population data per neighbourhood in Manhattan [33] to illustrate the approach of replacing the cost surface in emergency situations. The route to the emergency will be straight in order to save time, so the catchment areas for each emergency landing site are presented in Fig. 3.5. Each

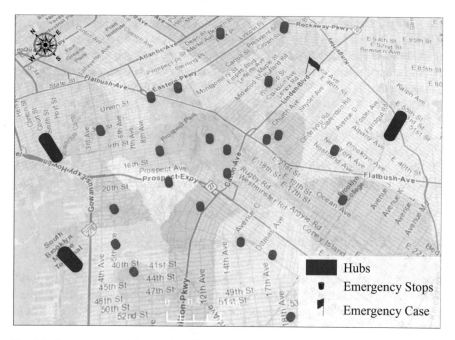

Fig. 3.5 Emergency handling with catchment areas based upon census data

colour represents one catchment area of one emergency stop. If the emergency case (red flag) happens, the corresponding landing site will be approached. The computation of the catchment areas depends on the distance to each emergency landing site and also the population density. The cost of the population density has an effect on the size of the catchment area, and it decreases the probability to cross areas with high population density. In reality, the cost surface changes permanently with the population density. This data is not common to be freely available. The data needs to be derived and estimated from the public transportation. Other possible data sources are georeferenced phone activities or data derived from surveillance cameras [29, 31, 34].

3.4   Optimize the Network Structure

The airspace needs to be optimized to maximize the potential for "greater efficiency and increased accessibility within transportation demand [with] increasingly flexible and scalable solutions" [9]. Possibilities to increase the efficiency within the usable aerial space include optimizing the network structure. This can be an adaptation of the static network structure to a time-depending network structure [35], but also the planning of the network's nodes. The site planning [7, 36] of the hubs and ports is crucial for the optimization of the network, as it directly influences the path lengths and the noisiness of the area. If a hub is planned right next to a school, it is impossible to compute a route that does not increase the average noise level at the school. Subsequently, the network structure should be planned taking into consideration variables like the average noise level, the average number of customers, the level of traffic congestion, the distance to the public transportation or the existence of emergency landing sites in a predefined radius. Most of the listed variables are time dependent, they change during the time of the day as well as during the day of the week and probably also during the week of the year. This requires a time depending network structure that is adaptable to the changing variables. This sub-chapter explains how we model the adaptable network considering changing conditions to increase the network's efficiency.

3.4.1 Time Depending Network Structure

The first changing condition regards temporal restrictions. These temporal restrictions can be modelled with dynamic geofences [8]. Events that trigger the temporal restrictions can be the presence of construction sites, open-air concerts, large assemblages of people, public speeches or any other kind of event that triggers higher safety requirements for the areas being used [17]. Dense fog or clouds are natural danger zones for aircraft navigation [37].

In the model, the dynamic geofences are combined with the static geofences and the cost surface is recalculated. The consequence is the recalculation of the flight network with the least cost paths avoiding also the temporal restricted areas. Figure 3.6 shows the recalculated network, because the yellow coloured temporal geofence intersects with the network. The blue coloured path is the new connection in the network.

Moreover, the network needs to be scalable for different customer demands. In rush hours, some parts of the network might be congested, as the distance between aircrafts needs to be regulated [8]. The proposed network can be vertically extended if parts of the network face such a high demand, that the regulations force some aircrafts to wait on the ground. An additional network layer is visualized in yellow in Fig. 3.7. In the case that the maximal flight height is reached with the vertical extended layer, the route will be redirected to be on the left or right side of the existing route at that route segment.

Furthermore, the market strategy of the air taxi companies requires on-demand transportation [1–3]. To this point, we did not point out how the model takes this requirement into account. The on-demand extension of the network is visualized in orange in Fig. 3.7 and illustrates how we embedded this need in the model. The idea of our approach is to enable on-demand flights without affecting the citizens with additional routes. Therefore, the cost surface is manipulated in order to force the extension to use the already existing network. This is achieved by reducing the cost at cost surface, where the earlier calculated routes are located. This manipulation increases the probability that the route uses the existing network.

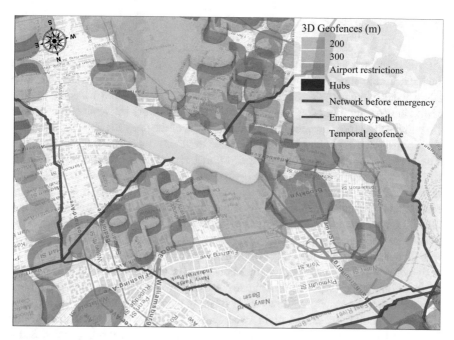

Fig. 3.6 Network comparison of least costs distance and noise

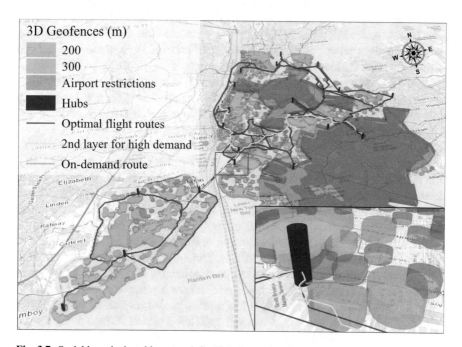

Fig. 3.7 Scalable and adaptable network for high-demand and on-demand situations

3.4.2   Site Planning for Hubs and Ports

At this point, we will outline how the site planning for the hubs and ports can be included in our model to improve the network structure. Ports are designed to be static nodes of the network, and the sites are connected to the public transport system [38]. Hubs on the other contrary are private and temporally active sites for landing [3] and are therefore considered as dynamic nodes of the network. The requirements for a port include first of all a close distance to other public transport possibilities. The ports need to be constructed close to airports, bus stations and train stations for becoming a competitive alternative to other transportations and to relieve the urban transport system [39]. The difficulties for the site planning occur with the legal restrictions mentioned in Sect. 3.2. Around airports, for example, restrictions do not allow the aircrafts to be manoeuvred close to the airport. Consequently, concepts need to be designed, which incorporate exceptional routes for these aircrafts, similar to helicopter routes in urban areas. The hubs are private ports and can be either an active or passive node in the dynamic network. As the flying objects can land on surfaces that are bigger than 392 square metres [30], the aircrafts can land on top of buildings if the structure allows it. Performing the structure analysis for all roofs of potential buildings can be highly demanding regarding the computation effort. However, there is a key factor in determining those suitable rooftops, which is the shape of the roof. Therefore, involving in this analysis, 3D point clouds from Light Detection and Ranging allows to determine in great detail the degree of suitability of the roof. Rooftops footprint data and LiDAR points from NY Open Data [33] have been used for this analysis. We decided to choose three different rooftops near the existing flight network with high levels of traffic noise. Each rooftop is evaluated with the height distributions of the LiDAR points of the rooftops. In this order of ideas, the first step was to generate or select the LiDAR data that lie within the edge of each rooftop footprint. Then, for each of the scenarios, we generate a box plot in order to assess the height distribution of the 3D cloud points for each building and a possible uniformity in the heights as illustrated in Fig. 3.8.

Thus, we can observe that the heights of the roofs in buildings 1 and 2 show a narrower interquartile range (IR) than those presented on rooftop 3. Statistically, this indicates higher uniformity in the heights of roofs 1 and 2. In other words, these roofs would have greater potential to develop hubs for this type of aircraft. Nevertheless, roof 2 has a slightly greater IR than roof 1. In addition, the range of roof 2 has lower heights than that of roof 1. Therefore, building 1 would be statistically the most suitable for building these structures. But uniformity could not only be guaranteed by a statistical analysis. Three-dimensional visualizations help to interpreting the suitability of a rooftop, as visualized in Fig. 3.8. This allows us to confirm that roof 1 has a roof with more uniformity or in other words more segments with flat surfaces. In contrast, roof 3 has a cross-gabled roof shape, which is not suitable for hubs development. However, building 2 shows different roof levels, which gives indications of the need to apply further techniques for the detection of patterns by minimum area required in the construction of these structures [38].

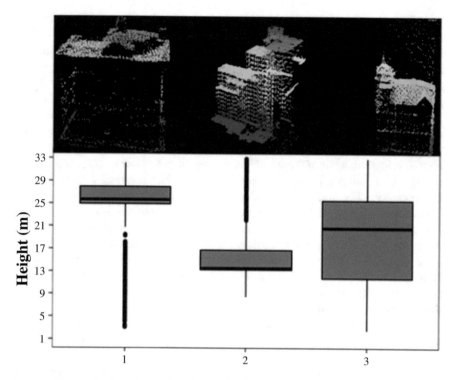

Fig. 3.8 LiDAR clouds with the points' height distribution in a box plot

In addition to massively developing this analysis to each of the roofs of New York City under the aforementioned conditions, a machine-learning approach would be very useful for detecting edges and objects from the LiDAR data.

3.5 Discussion

The presented results cover the requirements of the concepts from the flight agencies for dividing the future air space [8] while including additional concepts to get the consent and acceptance of the citizens for the flight routes. The air space is restricted due to the local laws with static geofences [14]. Local points and areas of interests can be protected with geofences that are adaptable with the parameters of the vertical and horizontal restrictions by the responsible authorities. These results for the study area of Manhattan are illustrated in Figs. 3.1 and 3.2 for the vertical and horizontal restrictions in Table 3.1. The proposed model includes also an option to produce a route network that takes into consideration the additional noise pollution. Following the guidelines of the World Health Organization [24] and concepts or organizations that deal with the environmental impact and the level annoyance with excessive noise pollution [26], we proposed a scalable solution to take the noise level into consideration. This is achieved by manipulating the cost surface in such a way that the resulting least cost network tends to follow areas that are already exposed to high noise from air and road traffic. The surface manipulation and the resulting network are illustrated in Figs. 3.3 and 3.4. By using a similar approach, an additional important safety aspect for the future urban air mobility [6, 8]. The underlying population density can be used to minimize the danger in cases of emergency. Our model includes catchment areas in a case of emergency, where the criteria risk minimization replaces the other criteria for the routes. If the aircraft is located in one of these catchment areas in an emergency case, as presented in Fig. 3.5, it follows the shortest possible path while avoiding high population densities [29]. The size of the catchment areas for each emergency stop is time dependent, as the population density varies by the time of the day and week [32]. By modelling the network as an adaptable and scalable network [8, 9], it is possible to include other time dependent variables in the network. One example are temporal impedances, and another one are additional vertical network extensions and on-demand routes, which are illustrated in Figs. 3.6 and 3.7. These time dependent network adaptations with temporal network nodes and edges allow the reaction to different needs and demands from customers (Fig. 3.8).

The combination of all the results in a parametrizable model is important to adapt to different situations that are time and location dependent. Different cities with different city structures logically need different air space restrictions to maintain or even improve the quality of live for the citizens by introducing the new green mode of transportation, and even for the same city, the requirements for the network change with the time. The model allows to incorporate the different conditions by manipulating the cost surface, which is used for the calculation of the least cost network.

It is important to point out that the available data for New York is very complete. The data for the ground surface, the flight paths of the planes and the landuse information are available for most of the big cities in the world, whereas the rooftops' heights or the traffic noise level is not common. The only data that needs to be processed beforehand in New York is the estimated population density per time.

Furthermore, the methodology of using a manipulative cost surface results in a shortcoming in the energy efficiency for following the routes. The reason for this is first of all the raster format of the cost surface. The raster format consists of cells that are squares. Therefore, the routes following the cells are not smooth and they have corners. Physically conditioned, the aircrafts would have to decelerate to zero at each corner if the paths were to be followed exactly. Another shortcoming of the cost surface is the fact that it is a plane and not a three-dimensional geometric body. A path might be more energy efficient if it was not following the minimal flight height, which would practically lead to a higher number of ascents and descents of the aircraft. The computed least cost network is therefore the basis for an optimization that takes into account the flight physics of the aircrafts. The aircraft specific flight behaviour in corners and the energy emissions in ascents and descents can be used in combination with the existing network for further optimization of the network. One suggestion for future work on this topic is the network optimization for the additional criteria. Another suggestion for future work is the further elaboration of the site selection for hubs and ports for additional criteria. The estimation and forecasting of the demand and the noise for each hour of a year could be used for selecting the optimal static and dynamic nodes in the network structure.

3.6   Conclusion

Networks for different criteria can be computed with the presented model for the urban air mobility of the future. The parametrized model is a tool that can be used by the responsible authorities or it can be used as consultation for regional planners to manage the future urban air space. The air space can be divided into restricted air space and non-restricted air space. The restricted air space complies with the local flight law, while the points and areas of interest like schools, universities or cemeteries can be protected. This is a possibility to avoid negative side effects that can possibly lead to a disapproval of the citizens in the particular urban area. The structure of the network is dynamic and adaptable to fit different purposes in different conditions that change with time. The criteria for the network calculation were the shortest distance, the noise and the least risk path in cases of emergencies. Furthermore, the features on-demand transportation and the scalable network extensions allow greater efficiency of the new mode of green transportation. In conclusion, the optimal network routes can be calculated for different purposes and scenarios for specific criteria. The network structure was optimized by analyzing the suitability for the hubs and ports.

This work proposes the practical approach for managing the urban air space of the future in Manhattan, but the model was built to be adaptable to all urban areas. The different requirements for different cities and different times can be met by running the model with different parameters, and the choice of the best fitting one can be done by the responsible authorities. The method enables urban planners, flight authorities and also the companies offering the flight service to investigate the flight routes and their negative side effects. The user of the model can define the minimal flight height and the boundaries of the restricted and protected air space with the vertical and horizontal 3D geofence height for each landuse or type of building. Except to defining these parameters to compute the restricted and protected air space, the user is also able to define the importance of the noise pollution for computing the network for the chosen parameters. The only requirement for running the model for other urban areas is the availability of the input data, which does not necessarily exist yet for all urban areas.

A recommendation for further improvement and completeness of the network is a pareto-optimization of the flight routes for more evaluation criteria of the network. Two of the variables that were not considered in the network were the energy consumption and time needed for different aircrafts to follow the network. These criteria cannot be included by manipulating the cost surface like noise; therefore, a different approach would be necessary to optimize the network. One solution to do this multi-criteria optimization in an appropriate time window might be the use of evolutionary algorithms.

References

1. Uber Elevate: Fast-forwarding to a future of on-demand urban air transportation: white paper (2016)
2. Lilium Aviation: Mission: aircraft for everyone (2019)
3. Volocopter: Reinventing urban mobility (2019)
4. Airbus: Urban air mobility: the sky is yours (2018)
5. Sean Captain: Uber's flying car chief on noise pollution and the future of sky taxis. Fast Company (2017)
6. National Aeronautics and Space Administration: Urban air mobility airspace integration concepts and considerations (2019)
7. Jiang, T., Geller, J., Ni, D., Collura, J.: Unmanned aircraft system traffic management: concept of operation and system architecture. Int. J. Transp. Sci. Technol. 5(3), 123–135 (2016)
8. Geister, D.: Integrating UAS into the future aviation system: a flexible approach enabling large-scale UAS operations: German aerospace center (DLR). Institute of Flight Guidance (2017)
9. Alexandrov, N.: Transportation network topologies: National Aeronautics and Space Administration (NASA). L-19047 (2004)
10. Federal Aviation Administration: Pilot's Handbook of Aeronautical Knowledge: 2016. Skyhorse Publishing, New York, NY, First Skyhorse Publishing Edition (2017)
11. European Union Aviation Safety Agency (ed.): Regulation of the European Parliament and of the Council of 4 July 2018 on common rules in the field of civil aviation and establishing a European Union Aviation Safety Agency, and amending Regulations. EASA, (eu) 2018/1139 edition (2018)
12. Civil Aviation Authority: General and special aviation rules (2018)
13. Dorr, L.: Finalize rules for small unmanned aircraft systems: press release. Federal Aviation Administration (FAA) (2016)
14. Federal Aviation Administration. Integration of civil unmanned aircraft systems (UAS) in the national airspace system (NAS) roadmap, first edition (2013)
15. Charles, A.: Case study Japan: GIS-technologies in urban planning: the future of urban development & services (2014)
16. Esri: Real-time data feeds and analytics: managing geofences (2018)
17. Hildemann, M., Delgado, C. (ed.) An adaptable and scalable least cost network for air-taxis in urban areas Study area: Manhattan, New York. In: AGILE, vol. 22 (2019). Accepted Short Papers and Posters
18. Open Data NYC: Open data of New York City: New York City population by neighborhood tabulation areas (2018)
19. Open Street Map: Open source data (2018)
20. Longley, P., Goodchild, M.F., Maguire, D., Rhind, D.W.: Geographic Information Science & Systems, 4th edn. Wiley, Hoboken, NJ (2015)
21. Dutta, S., Patra, D., Shankar, H., Alok Verma, P.: Development of GIS tool for the solution of minimum spanning tree problem using Prim's algorithm. In: ISPRS - International Archives of the Photogrammetry, Remote Sensing and Spatial Information Sciences, XL-8, pp. 1105–1114 (2014)
22. Civil Aviation Authority: Managing aviation noise (2014)

23. Sustainable Aviation: Sustainable aviation noise road-map (2019)
24. Geneva World Health Organization: Guidelines for community noise (1999)
25. 2018 AIAA/ASCE/AHS/ASC Structures, Structural Dynamics, and Materials Conference, Reston, Virginia, 01082018. American Institute of Aeronautics and Astronautics
26. U.S. Environmental Protection Agency. Office of noise abatement and control: EPA (1974)
27. National Transportation Noise Map. Exposure to noise from aviation and interstate highways (2014)
28. Hurtley, C.: Night Noise Guidelines for Europe. World Health Organization Europe, Copenhagen (2009)
29. Di Donato, P., Atkins, E.: Evaluating risk to people and property for aircraft emergency landing planning. J. Aerosp. Inf. Syst. **14**(5), 259–278 (2017)
30. Vaddi, S., Kumar, C., Jannesari, A.: Efficient object detection model for real-time UAV applications (2019)
31. Ayeni, A.O., Musah, A., Udofia, S.K.: Assessment of potential aerodrome obstacles on flight safety operations using GIS: a case of Murtala Muhammed International Airport, Lagos-Nigeria. J. Geograph. Inf. Syst. **10**(01), 1–24 (2018)
32. Fung, J.: Manhattan population explorer (2019)
33. Open Data NYC. Open data of New York City: building footprints (2018)
34. Douglass, R.W., Meyer, D.A., Ram, M., Rideout, D., Song, D.: High resolution population estimates from telecommunications data. EPJ Data Sci. **4**(1), 334 (2015)
35. Musliman, I., Rahman, A.A., Coors, V.: Implementing 3D network analysing in 3D-GIS: the international archives of the photogrammetry, remote sensing and spatial information sciences. Paper, Universiti Teknologi Malaysia, Skudai, Johor, Malaysia (2008)
36. Dueker, K., Delacy, B.: GIS in the land development planning process balancing the needs of land use planners and real estate developers. J. Am. Plann. Assoc. **56**(4), 483–491 (1990)
37. Fallows, J.: Free Flight: A New Age of Air Travel. Oxford Publicity Partnership, New York (2002)
38. Fadhil, D.N.: A GIS-based analysis for selecting ground infrastructure locations for urban air mobility. Master's Thesis, Munich, 2018. Master (2018)
39. Archer, J., Black, A., Roy, S.: Analyzing air taxi operations from a system-of-systems-perspective using agent-based modeling. Purdue University West Lafoyette (2012)

Chapter 4
Putting the SC in SCORE: Solar Car Optimized Route Estimation and Smart Cities

Mehrija Hasicic and Harun Siljak

Abstract Solar exposure of streets and parking spaces in dense urban areas varies significantly due to the infrastructure: buildings, parks, tunnels, multistorey car parks. This variability leaves space for both real-time and offline route and parking optimization for solar-powered vehicles. In this chapter we present Solar Car Optimized Route Estimation (SCORE), our optimization system based on historic and current solar radiance measurements. In addition to the comprehensive review of SCORE, we offer a new perspective on it by embedding it in the bigger picture of smart cities (SC): we analyze SCORE's relationship with the smart power generation and distribution systems (smart grid), novel transportation paradigms, and communication advancements. While the previous work on SCORE was focused on technical challenges which are described in the first part of this chapter (optimization, communication, sensor data collection, and fusion), here we proceed with a systemic approach and observe a SCORE-equipped unit in the near-future society, examine the sustainability of the model and possible business models based on it. We consider the problem of vehicle routing and congestion avoidance using incentives for users on non-critical journeys and co-existence of SCORE and non-SCORE using vehicles. Realistic pointers for SCORE-aware design of infrastructure are also given, both for improved data collection and improved solar exposure while considering trade-offs for non-SCORE users.

M. Hasicic (✉)
Electrical and Electronics Engineering Department, International Burch University Sarajevo, Ilidža, Bosnia and Herzegovina
e-mail: mehrija.hasicic@ibu.edu.ba

H. Siljak
CONNECT Centre, Trinity College, The University of Dublin, Dublin, Ireland
e-mail: harun.siljak@tcd.ie

© Springer Nature Switzerland AG 2020
H. Derbel et al. (eds.), *Modeling and Optimization in Green Logistics*,
https://doi.org/10.1007/978-3-030-45308-4_4

4.1 Introduction

When we introduced Solar Car Optimized Route Estimation (SCORE), we were
solving the problem of a single, possibly unique solar vehicle optimizing its route
in a *conventional* city. In this chapter we examine the role of such a smart system,
devised to find a sunny route (and a sunny parking lot) for a solar car, in a larger
smart ecosystem, one of a *smart* city.

We recognize that solar vehicles of different sizes and different types of
hybridization [1, 2] have been proposed, as well as options to make existing cars
hybrid solar vehicles [3], and that solar cars are a part of a larger scheme in the vision
of smart city, smart transport, and any future involving fighting climate change [4, 5].
How to fit SCORE-guided solar vehicles in a smart transportation system hosting
other electric and hybrid vehicles, how to fit it in a smart grid system which is
sensitive to electric cars being plugged in, and how do they fit in the landscape of
smart city's parking lots which should serve as charging stations? These are the
questions we share our opinions on, after presenting the idea of SCORE in brief.

4.2 SCORE

Solar Car Optimized Route Estimation (SCORE) system is a system that deals with
the data acquisition, software and hardware developed propositions for route and
parking selections for a (hybrid) solar vehicle. Detailed explanation of the system is
given in the following subsections.

4.2.1 System Description

The SCORE is built up from three separate parts shown in Fig. 4.1 and listed
below:

- irradiance sensing—mobile sensor that transmits solar radiance from the streets
  in real time. These transmitter should, preferably be placed on the frequently
  moving cars or could be placed on fixed locations in the streets;
- cloud servers—server for data fusion that collects the data from the field and
  the third party and combining it with offline data and sending it to the device
  embedded in the cars in form of matrix;
- cruise computer—embedded computer unit in the solar car that combines
  received data with its own readings and normalizes the data using built-in light
  sensor. Proposed route changes dynamically based on weather updates.

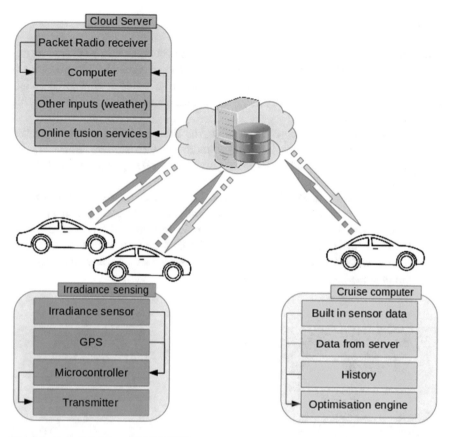

Fig. 4.1 Overall structure of SCORE [6]

Prototype implemented in this research uses only historical data from the cloud due to the limitations of self-built solar car, and due to the fact that only one car in this case uses SCORE.

4.2.2 Data Collection and Management

As previously mentioned SCORE is designed to work with both, online and offline data. Two ways of collecting it are proposed: weather forecast and mobile sensor installed on solar and regular cars (i.e., taxi, delivery tracks, etc.).

4.2.2.1 A Priori Data

Architecture of the device (using components compact enough to be placed on any car to collect the data without any needs for route or vehicle customization) is shown in Fig. 4.2a. If placed on taxis or delivery tracks this device could collect huge amounts of data random in time, date, and location, opening the space for big data analytics as well.

Proposed way in this research for sending the data is APRS (Automatic Packet Reporting System) as the easiest packet radio implementation. However, any wireless protocol can be used for the same purpose as well as proprietary and private frequencies.

The server, shown in Fig. 4.2b, receives the data and converts it to text using common sound card and appropriate software. This data and data from CAD (Computer Aided Design) and GIS (Geographic Information System) is put in tabular form online. CAD data is obtained by simulating sun movement in 3D model of the street, while GIS data is provided by GIS services measuring solar radiance of different areas.

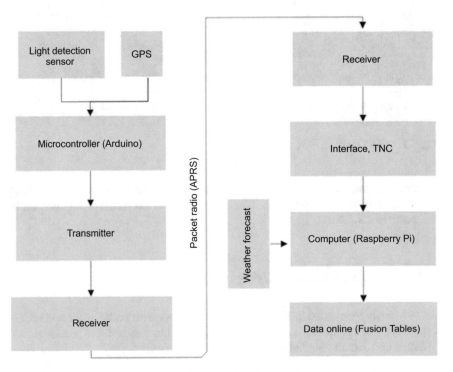

Fig. 4.2 (**a**) Mobile device structure collects the data from GPS and light detection sensor and transmits it to the (**b**) Central server structure which process the data and stores it online, available for end user (EV drivers) to download it [6]

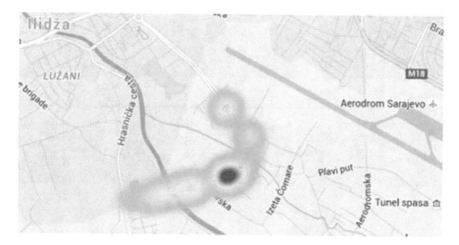

Fig. 4.3 Visualization of the data from Fusion Table for local map showing irradiation on the route from our University to the city center. Irradiation has been showed by the colors, red representing irradiation 1—very sunny route and green representing low irradiation—route being in the shadows of near buildings or trees [7]

Shown in Fig. 4.2 is an exemplary scheme of the technologies proposed in this research which can always be replaced by some existing alternatives. Fusion tables are proposed due to the ability of presenting the data visually as shown in Fig. 4.3.

4.2.2.2   Real-Time Data

I/O architecture of the embedded computer placed in the solar car is shown in Fig. 4.4. To optimize the route system uses routes from user's history, sensor fusion data from the cloud where the most resent data is the one that matters the most , measurements from the solar panels and battery, and measurement from the built-in light sensor.

Measurements from the built-in sensor are used as corrective measurements of the real output and predicted one, where error rate is used to calibrate all others predicted values. On the other hand, electrical measurements are used to predict energy consumption and cost of every available route from beginning to end destination. All the facts and measurements taken into consideration result in proposition of the best route and parking with high solar irradiation. Combination of the multiple objectives (route and parking) deserves more attention in future research, as well as variation in parameters of the optimization problem.

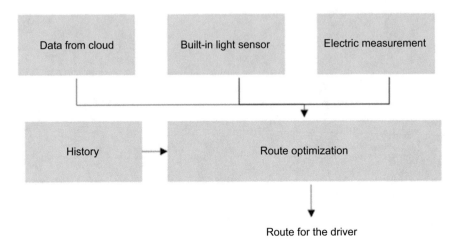

Fig. 4.4 Embedded car computer structure [7]

4.2.3 Theoretical Considerations

Theoretical considerations must be taken into account when considering two things: solar irradiation data and optimization. Since we have proposed two types of data collection each has its own considerations to be analyzed: Online data—normalized data represented from 0 (no radiance) to 1 (maximum radiance) with a time-stamp denoting the time expressed in hours with beginning of the year as the reference point; Offline data—data predicted for particular area using weather forecast, CAD, and GIS. Final irradiation value, r, is calculated using normalized irradiation values from previously mentioned online and offline data, r_{on} and r_{off}, expressed as:

$$r = r_{on} \cdot a + r_{off} \cdot (1 - a) \qquad (4.1)$$

where a is expressed as:

$$a = \exp\left(\frac{-(t_{curr} - t_{meas})^2}{100000}\right) \qquad (4.2)$$

and t_{curr} and t_{meas} are time expressed in hours from the beginning of the year, where t_{curr} stands for current time and t_{meas} represents the time of last measurement of the data. Denominator is chosen empirically.

Dijkstra's optimization algorithm is chosen for the route optimization where positive weights of edges and nodes needed to be decided. All main crossroads in Sarajevo were chosen to be nodes. Beside the length of the road which is obvious factor, we also need to take into consideration converted solar energy while on the road. Car specifications that affect the converted solar energy are:

- motor power—11 kW
- panel area—$2 \times 0.726 \, m^2$
- panel efficiency—18%
- received power per square meter of the panel—957 W/m^2 under maximum radiance, which results in 30% of radiance is reflected.

In this research for the purpose of developing prototype we were doing calculations and server serves only one car at the time. The idea is to develop it further more to be functional for unlimited number of solar cars at the same time. The figures in the specification are relevant as they define the efficiency of solar energy conversion and the requirements for the operation of the vehicle. With greater conversion efficiency expected in new car development, SCORE's effect will be augmented as well.

Solar irradiation of the road segments is taken as the arithmetic mean of values taken from beginning and end node. This crude approximation is used for the prototype development but will be avoided in case when mobile sensors are installed on cars and start cruising the city.

As for the parking lot, both irradiation of the parking and its distance from the destination are taken into consideration. Parking with the highest ratio of irradiation and distance is the one that is going to be proposed to the user. Whether irradiation or the distance is more important can be decided by user by taking either one or another to a non-unit power.

All previously mentioned calculations are done on server and client only gets the proposed route and parking. However maps are updated to reflect changes in reality and update of calculated routes is regularly conducted since weather may vary significantly, especially on longer trips. Frequency of updates can be variable (e.g., urban trips vs. intercity trips).

4.2.4   Implementation

4.2.4.1   Implementation of Sensor Data Collection and the Server

As previously mentioned, mobile device developed in the scope of this research is compact and autonomous which allows its placement on the car without customization of the car itself. For the data collection purpose it can be placed on any vehicle, not only on the solar cars.

APRS is proposed as transmission channel for delivering GPS position and sensor data to the terminal node. It has been used in monitoring systems for the same purpose [8]. For the transmission we have used amateur radio bands; however, different frequencies and wireless protocols can be used for the different implementation of the SCORE.

Since we are interested not only in historic data but also in the real-time data we need to be able to constantly update and communicate with the server.

We did not need large computing power for the server so we implemented it on Raspberry Pi 2, even so the same could be done on cloud as well. Enabling all parties to access fusion tables was the main reason to store them on the cloud, this allows client to access them on their smart phones and/or computers as well.

As the SCORE system scales up and more vehicles adopt it, various approaches for congestion avoidance can be employed: caching on a cloud-edge scheme, local broadcasting, etc.

4.2.4.2   Implementation of the Optimization Client

Beside previously mentioned prototype done on microcontroller we have also implemented testbed for algorithm testing in MATLAB (simulation of the server–client communication). Graphical user interface (GUI) in MATLAB is shown in Fig. 4.5.

As it was previously mentioned, in the beginning stage of our research we have planned to compute car computer on ARDUINO. However due to the significantly more memory, needed to keep whole matrix representing the graph in working memory of the processor, the car computer prototype was built on TI's ARM Cortex-M4F based TM4C123G LaunchPad.

The car computer has a keyboard, used for entering the destination by choosing the node number, and a display used for showing the sequence of nodes (crossroads) the client needs to traverse in a particular order. Simple implementation we did, shown in Fig. 4.6, gets the matrix data through wireless or USB debug cable and

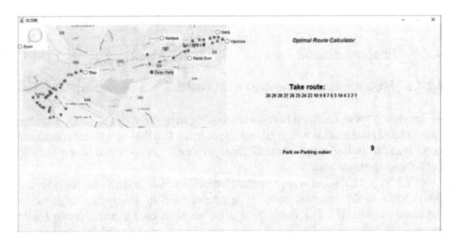

Fig. 4.5 GUI in MATLAB. Each red dot is a node that has been assigned a number. Using those numbers computer shows us the optimum route. Data with parking slots near every node is provided to the software, based on equations it calculates and shows you nearest parking with highest irradiation too

Fig. 4.6 The navigation
client placed in the car [9]

optimizes using Dijkstra's algorithm and displays the set of nodes connecting
starting and end node.

4.3   Putting the Smart City (SC) into SCORE

Solar energy resources are, alongside other renewables, instrumental element of
smart cities [10]. However, the typical solar power discourse in smart cities is one
of homes and power stations, fixed objects with a fixed photovoltaic setup. What
changes when we include solar cars into the equation?

It is in the case of a single solar vehicle using SCORE, not much changes in the
smart city: one driver will use routes and parking lots recommended by SCORE,
while the only effect that can be felt in community is one coming from the size of
the mobile sensor—community updating solar irradiance data online (which, as we
have already noted, would be independent from solar cars themselves, and would be
placed on vehicles performing public services).

It is much more interesting to examine the situation when a growing community
of drivers uses SCORE: a significant portion of population uses hybrid solar cars,
alongside those who use the non-solar hybrid cars, and maybe a portion of fossil fuel
cars as well in the transitional period. In this scenario, the question of other vehicles
occupying best routes for solar cars may be a hindering issue (and a frustrating
one for solar drivers). This is where SCORE needs integration with other software
and hardware platforms in the future vehicular networks: be it in our contemporary
navigation, mapping, and route recommendation systems, or in the autonomous
driving systems of the future, SCORE input would give recommendations to non-
SCORE drivers to use other routes when SCORE drivers are expected to use their
well-insulated routes. Direct incentive is jam avoidance, but a different in-traffic
economy could be built as well, with incentives like good parking spots or free

charging station time for cars that choose to go with less attractive routes, to balance the traffic.

This is what the citizens of the smart city (at least those who drive) see as the effect of introduction of SCORE. What is the effect noticed by sensors in the power grid and the dispatchers in the smart grid of the smart city? SCORE-enabled vehicles, being at least partially solar, would operate as gridable elements, contributing power into the system or taking it out of the system while plugged in, depending on the battery state. The effect of gridable vehicles is studied in literature [11, 12], but the vehicles generally observed are those who charge solely from the grid, i.e., not having own renewable source as solar vehicles do. An additional renewable source connected to the grid can offer a new degree of freedom for the smart grid. The idea behind smart parking for electric vehicles revolves around smart power management of cars connected to grid [13]. If the solar component is included, it is the parking lot itself with solar panels [14], but here we bring the solar panels to places that need them (and places that can use them well based on their insolation), encouraging the drivers to park there.

In both of these scenarios, it is the relationship between a SCORE-user and a non-SCORE entity that determines the success of implementation.

A SCORE-user wants a (1) satisfying travel and (2) convenient parking. The adjectives used here are vague: a satisfying travel could be the one that is as short as possible (if that is the priority) or the one that is as cheap as possible (if that is the priority), or a mix of the two. Similarly, the desired parking lot could be the most economical one (charging the most in high insolation conditions) or simply the most comfortable (i.e., closest to the destination). From a game-theoretic perspective, the question is how conflicting these interests of the SCORE-user are with those of other, non-SCORE users.

An electric vehicle (EV) driver could state the same vague requests for travel and parking as our SCORE-user, but the mechanics behind them would be different, as the EV driver's power budget can never increase "for free." This simplifies the optimization for the EV driver, as they aim to minimize the losses, not able to expect a power contribution from the outside. This in practical conditions means that often the optimal routes for SCORE and non-SCORE drivers will not be the same, and they do not have to directly compete for them. In cases where they do compete, the solution can be brought by either congestion pricing [15] or elaborate pricing schemes that have been developed for electric vehicles [16].

The other non-SCORE entity in the game is the grid, and it plays the same game with the EV driver as it does with a SCORE driver, with the difference that a SCORE vehicle would serve as a mobile renewable source if parked for a long time period, a benefit not seen from the regular EV (which at best would serve as a mobile battery). If we extend this player's scope to include the general infrastructure, it can include the parking lot planning—and insightful planning of parking lots with natural sunlight would come as a new factor, opposing the previous architectural desire to create as much shade as possible in parking spaces [17].

4.4   Conclusions

SCORE is a convenient framework for a small group of solar vehicles finding their way in a modern town. Only with the smart, connected cities of the future will it see its full potential integrating into smart transport and smart grid of the city alike. A social benefit to the ecosystem and providing the service to its user are key benefits of implementing the SCORE into SC. We argue that it will complement other smart transport solutions in a natural fashion, without disruptions and deadlocks in optimization.

SCORE's promise of integration lies in its dual flexibility and robustness: it is responsive as it tracks all relevant changes in the environment, and with its multi-criterial extensions (parking, route, etc.) it can provide a wide spectrum of solutions catering for different needs (criticality, extreme pollution decrease, daily congestion management).

Acknowledgments   We thank Mr Damir Bilic (Mälardalen University, Sweden) for his invaluable contribution in developing SCORE, and useful discussions in preparation of this chapter.

References

1. Mulhall, P., Lukic, S.M., Wirasingha, S.G., Lee, Y., Emadi, A.: Solar-assisted electric auto rickshaw three-wheeler. IEEE Trans. Veh. Technol. **59**(5), 2298–2307 (2010)
2. Arsie, I., Rizzo, G., Sorrentino, M.: Optimal design and dynamic simulation of a hybrid solar vehicle. SAE Technical Paper 2006-01-2997, SAE International, Warrendale (2006)
3. de Luca, S., Di Pace, R.: Aftermarket vehicle hybridization: potential market penetration and environmental benefits of a hybrid-solar kit. Int. J. Sustain. Transp. **12**(5), 353–366 (2018)
4. Popiolek, N., Thais, F.: Multi-criteria analysis of innovation policies in favour of solar mobility in France by 2030. Energy Policy **97**, 202–219 (2016)
5. Joshi, M., Vaidya, A., Deshmukh, M.: Sustainable transport solutions for the concept of smart city. In: Gautam, A., De, S., Dhar, A., Gupta, J.G., Pandey, A. (eds.), Sustainable Energy and Transportation: Technologies and Policy, Energy, Environment, and Sustainability, pp. 21–42. Springer, Singapore (2018)
6. Hasicic, M., Bilic, D., Siljak, H.: Criteria for solar car optimized route estimation. Microprocess. Microsyst. **51**, 289–296 (2017)
7. Hasicic, M., Bilic, D., Siljak, H.: Sensor fusion for solar car route optimization. In: 2016 5th Mediterranean Conference on Embedded Computing (MECO), pp. 456–459 (2016)
8. Bello, A.G., Torres, D.A.A.: Design and construction of an agrometeorological monitoring system using APRS. Revista Colombiana de Tecnologías de Avanzada **1**(9), 127–132 (2007)
9. Bilic, D., Hasicic, M., Siljak, H.: Practical implementation of solar car optimized route estimation. In: 2016 XI International Symposium on Telecommunications (BIHTEL, pp. 1–5 (2016)
10. Wang, S., Wang, X., Wang, Z.L., Yang, Y.: Efficient scavenging of solar and wind energies in a smart city. ACS Nano **10**(6), 5696–5700 (2016)
11. Su, W., Eichi, H., Zeng, W., Chow, M.: A survey on the electrification of transportation in a smart grid environment. IEEE Trans. Ind. Inf. **8**(1), 1–10 (2012)
12. Saber, A.Y., Venayagamoorthy, G.K.: Efficient utilization of renewable energy sources by gridable vehicles in cyber-physical energy systems. IEEE Syst. J. **4**(3), 285–294 (2010)

13. Honarmand, M., Zakariazadeh, A., Jadid, S.: Optimal scheduling of electric vehicles in an intelligent parking lot considering vehicle-to-grid concept and battery condition. Energy **65**, 572–579 (2014)
14. Nunes, P., Figueiredo, R., Brito, M.C.: The use of parking lots to solar-charge electric vehicles. Renew. Sustain. Energy Rev. **66**, 679–693 (2016)
15. Washburn, D., Sindhu, U., Balaouras, S., Dines, R.A., Hayes, N., Nelson, L.E.: Helping CIOs understand "smart city" initiatives. Growth **17**(2), 1–17 (2009)
16. Alizadeh, M., Wai, H.-T., Chowdhury, M., Goldsmith, A., Scaglione, A., Javidi, T.: Optimal pricing to manage electric vehicles in coupled power and transportation networks. IEEE Trans. Control Netw. Syst. **4**(4), 863–875 (2016)
17. McPherson, E.G.: Sacramento's parking lot shading ordinance: environmental and economic costs of compliance. Landscape Urban Plan. **57**(2), 105–123 (2001)

Chapter 5
Evaluation and Prioritisation of Green Logistics and Transportation Practices Used in the Freight Transport Industry

Aalok Kumar and Ramesh Anbanandam

Abstract The freight transportation is a potential source of environmental sustainability degradation and having negative impact of supply chain performance. To address several negative impacts of freight logistical activities, this chapter develops a multi-criteria decision-making (MCDM) framework for evaluating green freight transport practices (GFTP) used by the freight transport companies in emerging economy, i.e., India. The importance of GFTPs is computed with the help of fuzzy best–worst method (FBWM). This chapter presents the ranking of the GFTPs based on the linguistics input given by the industry expert. The proposed framework is validated with a case of the Indian freight transport industry. The FBWM based analysis reported that competitive pressure from other freight transport companies is an important factor for adopting green transport practices. The model robustness checked with the sensitivity analysis and rank variation is presented. The specific managerial implications of the research would help logistics managers in terms of promoting green freight transport in emerging economies.

5.1 Introduction

Freight transportation is the largest source of emissions around the world. The inclusion of green logistics and transport practices is essential to address the sustainable development of the freight transport industry [1]. The sustainable freight transportation system will help to improve the economic and social well-being of the local communities [2]. World Commission on Environment and Development [3] reported that to achieve the economic growth of organisations, the inclusion of environmental and social sustainability measures is required. The international policybodies such as the United Nation's Organisations, World Business Council for Sustainable Development, and the National Government are focusing on green

A. Kumar (✉) · R. Anbanandam
Department of Management Studies, Indian Institute of Technology, Roorkee, India
e-mail: ramesh.anbanandam@ms.iitr.ac.in

© Springer Nature Switzerland AG 2020
H. Derbel et al. (eds.), *Modeling and Optimization in Green Logistics*,
https://doi.org/10.1007/978-3-030-45308-4_5

transport practices to reduce the transport externalities. The transport environment externalities are related to the CO2 emission, NOx emission, congestion, accidents, and consumption of fossil fuels [1]. The freight transport sector consumes one-fifth of the global energy demand [4] and is responsible for one-fourth to one-half of the total logistics cost.

The freight transport sector is the big consumer of global energy deamd, it causes several climate change externalities on the earth planet. The increased global climate change issues have taken the attention of researchers and policymakers to consider environmental sustainability measure framework as a lever to achieve the sustainability of the freight transport industry. Liu et al. [5] highlight the importance of environmental sustainability research in the freight transport industry, and many authors highlight the studies to address the freight transport externalities [1, 6–9].

Despite the past studies discussing the environmental impact of manufacturing organisations, a little focus has been given to the logistics industry [10]. In order to improve the sustainability performance of the freight transport industry, organisations should incorporate the environmental sustainability practices in their supply chain operations [11].

The following research questions are addressed through this research:

- What is the important green freight transport practices for improving the environmental sustainability of the emerging economy freight industry?
- How GFTPs are prioritised for necessary policy deployment?
- How to compute the priority weight of GFTPs?

The above research questions are addressed through the following objectives:

1. to identify the important green freight transport practices for improving the environmental sustainability of the freight logistics industry;
2. to compute the importance weight of green freight transport practices (GFTP) by using a novel fuzzy best–worst method (FBWM) method;
3. to prioritise the GFTPs and test the model robustness through sensitivity;
4. to put forward the managerial implication of the research.

The above objectives are achieved in the following manner. First, the important green freight transport practices are identified from the literature review and their relevance is obtained through industry experts. The proposed framework is validated with the case of the Indian freight industry. Second, the final list of GFTPs is analysed with the help of a multi-criteria decision-making method, i.e., fuzzy best–worst method (FBWM). The rationale of using the FBWM method is that this method required less number of pairwise comparisons and improved consistency compared to analytic hierarchy process (AHP) and other MCDM methods [12]. Third, the importance weight of each GFTP is prioritised, and a sensitivity analysis is performed for checking model robustness. The findings of this chapter would help to improve the greenness of the freight transport industry. The industry can use the proposed framework for designing the organisation's green strategy for green performance measures.

The remainder of the chapter is organised in the following manner. Section 5.2 presents the background of the research area and the identification of important green practices used in the freight transport industry. The proposed solution methods steps are presented in Sect. 5.3. The application of the proposed framework is given in Sect. 5.4. The result discussion and sensitivity analysis are presented in Sect. 5.5. Finally, Sects. 5.6 and 5.7 present the managerial implications and conclusion of the research.

5.2   Literature Review

In logistics activities, transport has the largest environmental impact [13]. Freight shippers and logistics service providers (LSPs) are key players of the freight transport industry and need to support green supply chain management (GSCM) in the freight transport industry [14]. Freight transport organisations in developed countries face huge pressure to adopt green transport practices [15]. They also reported that GSCM had not been adopted very widely in logistics sector and their green benefits are limited [14]. Björklund [16] highlighted the importance of environmental responsible purchase behaviour in transport service selection. The author highlighted internal management practices, green image, resources of the firm, and governmental environmental policies as an important driver for adopting green practices. The internal green practices significantly contribute to the overall sustainability performance [8]. They reported that support of top management for green activities is necessary for adopting green practices in logistics industry. The top management plays a pivotal role to implement green practices, the adoption of newer and greener technologies, and various green awareness programs among the employees. The role of top management in the development of environmentally sustainable practices in third-party logistics organisation (3PL) has been presented by Lin and Ho [10]. Lieb and Lieb [17] reported that sustainability goal formation, intercompany sustainability assessment groups, environmental checklists, and publication of sustainability reports are few initiatives to improve the green performance of the freight transport industry. Liesen et al.'s [18] research indicated that freight companies might report GHG emissions and other environmental activities and share that information with stakeholders. Colicchia et al. [8] concluded that intra-organisational and inter-organisational green initiatives are required to improve the environmental sustainability of LSPs operations. Affi et al. [19] developed a variable search algorithm for the green vehicle routing to improve the environmental sustainability of the freight transportation. Kumar and Anbanandam's [20] literature review highlighted the importance of environmental and social sustainability considerations in freight transport network design for improving sustainable performance.

Collaborative actions to improve the environmental sustainability of the supply chain could slow down in the absence of standard methodologies that would allow companies to measure environmental impacts and share the costs and benefits of environmental initiatives [8, 21]. Ramanathan et al.'s [22] study showed the impact of collaboration in reducing GHG emission in the freight transport industry. Furthermore, freight shippers are rarely conscious of environmental sustainability; LSPs might guide them to improve the sustainability of freight distribution [23]. Freight shippers can make decisions while purchasing transport services, and LSPs play a pivotal role in developing environmentally sustainable freight transport solutions [8, 24]. Sureeyatanapas et al. [25] reported that the government policy and regulations, competitive pressure from other firms' green practices, and eco-driving practices are major drivers for green practices adoption in LSPs domain.

The LSPs should adopt a multimodal transport system for long hauling [26]. They claimed that a modal shift from road mode to intermodal mode significantly reduces the amount of emission and congestion [27]. Lammgård's [28] study shows that intermodality in freight transportation would reduce CO_2 emissions and improve environmental performance. Wong et al. [29] developed an optimisation model for truckload operations to reduce the CO_2 emission and recommended for the full load transport. Liimatainen et al. [30] developed a decision support tool for assessing the environmental performance of Nordic countries logistics service providers through energy efficiency index. They recommend that full vehicle loading, reduce empty running, and route optimisation are a winning factor for environmental sustainability. Perotti et al. [15] show that various environmental initiatives associated with the individual organisations' green practices such as eco-driving, reducing empty haul of the vehicle, full vehicle loading, and route optimisation techniques [25].

The environmental awareness and training of the employees would help to improve the environmental performance of the firm [31, 32]. The organisation support for developing green transport practices is essential [17]. Some of the transport companies considered that development of formal green/environmental sustainability goal statement that lead to environmental sustainability initiatives [7]. The organisational employees need sufficient knowledge about environmental sustainability practices to meet the customers (shippers) requirement [28]. Colicchia et al.'s [8] work focuses on internal human resource management practices of logistics service providers. They recommended that environmental training programmes at all levels have key importance in the achievement of companies' sustainability goals. Bask and Rajahonka [33] highlighted that government subsidies or tax breaks might sometimes support the adoption of such environmental practices. The above reviewed studies clearly show that there is a lack of green freight transport practices framework for improving the green performance of the freight transport industry. The final list of important green freight transport practices is given in Table 5.1.

Table 5.1 Important GFTPs of freight transport industry

#	Green freight transport practices (GFTPs)	References
1	Support form top management on green practices (GFTP1)	[17]
2	Assessment of GHG emission and annual sustainability report publication (GFTP2)	[8, 14]
3	Environmental management system certification (GFTP3)	[33, 36]
4	Promoting collaborative transport practices with other transport organisations (GFTP4)	Expert opinion
5	Competitive pressure from other freight transport firms' green practices (GFTP5)	[17, 25]
6	Promoting multimodal services for long hauling (GFTP6)	[26, 28]
7	Reduction in empty vehicle run (GFTP7)	[8, 25]
8	Environmental sustainability knowledge and awareness programs (GFTP8)	[31, 32]
9	Incentives green freight transport practices (GFTP9)	[25, 36]

5.3   Method

This section presents the computational step of FBWM. A best–worst method is a novel MCDM method developed by Professor Razaei [12]. The flow chart of used methodology is given in Fig. 5.1. This method requires lesser amount of pairwise

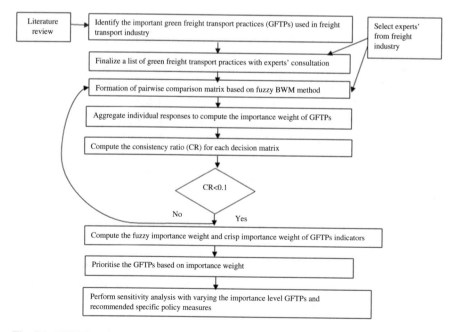

Fig. 5.1 FBWM methodology flow chart

comparison than other MCDM methods [12, 38]. The extension of the BWM method into fuzzy environment helps to incorporate the decision-making ambiguity and imprecisions of the decision-making process [34]. This chapter uses triangular fuzzy number (TFN) for collecting the experts opinion are measured on triangular fuzzy scale [35]. The FBWM method has the following steps:

Step 1: Develop a hierarchical structure of the decision-making problem.
In this step, the selection criteria have been identified. Let us assume $\{c_1, c_2, ...c_j....c_n\}$ are the criterion set of n criterion.

Step 2: Identification of most preferred, most important, or most favourable criteria and the poorest performing, least important, and least favourable criteria. The above identification gives the information of best and worst vector, represented as cB and cw, respectively.

Step 3: After identification of cB and cw, the pairwise distance vector from best and worst vector are identified on linguistics scale presented as "Equally preferred (EP)", "Weakly preferred (WP)", "Fairly preferred (FP)", "Strongly preferred (SP)", and "Absolutely preferred (AP)" [41]. Based on the above scale, best-to-other (BO) and others-to-worst (OW) vectors can be determined. The corresponding triangular fuzzy values of linguistics scale are as follows: (1,1,1), (2/3,1,3/2), (3/2.2,5/2), (5/2,3,7/2), and (7/2, 4, 9/2) [41]. The corresponding distance vectors are identified as $\tilde{A}_{BO}$ and $\tilde{A}_{OW}$ computed with Eqs. (5.1) and (5.2).

$$\tilde{A}_{BO} = (\tilde{a}_{B1}, \tilde{a}_{B2}......\tilde{a}_{Bj}.....\tilde{a}_{Bn}) \qquad (5.1)$$

$$\tilde{A}_{OW} = (\tilde{a}_{1W}, \tilde{a}_{2W}......\tilde{a}_{jW}.....\tilde{a}_{nW}) \qquad (5.2)$$

where $\tilde{a}_{Bj} = (a_{Bj}^L, a_{Bj}^M, a_{Bj}^U)$, and in the case of $j = B$ (best criteria), then $\tilde{a}_{Bj} = (1, 1, 1)$, and when $j = W$ (worst criteria), then $\tilde{a}_{jW} = (1, 1, 1)$.

Step 4: Determine the optimal weight of n criteria, i.e., $(\tilde{w}_1, \tilde{w}_2, ...\tilde{w}...\tilde{w}_n)$. The optimal weight of each indicator for each compared fuzzy pair $\frac{\tilde{w}_B}{\tilde{w}_j}$ and $\frac{\tilde{w}_j}{\tilde{w}_W}$ should satisfy the following conditions given in Eqs. (5.3) and (5.4).

$$\frac{\tilde{w}_B}{\tilde{w}_j} = \tilde{a}_{Bj}(j = 1, 2, ...n) \qquad (5.3)$$

$$\frac{\tilde{w}_j}{\tilde{w}_W} = \tilde{a}_{jW}(j = 1, 2, ...n) \qquad (5.4)$$

The computed optimal weight should minimise the maximum gaps of $\left|\frac{\tilde{w}_B}{\tilde{w}_j} - \tilde{a}_{Bj}\right|$ and $\left|\frac{\tilde{w}_j}{\tilde{w}_W} - \tilde{a}_{jW}\right|$ for all j are minimum. The above-constrained optimisation programming is used to compute the fuzzy weight of the indicators:

$$\min_{j} \max \left\{ \left| \frac{\tilde{w}_B}{\tilde{w}_j} - \tilde{a}_{Bj} \right| , \left| \frac{\tilde{w}_j}{\tilde{w}_W} - \tilde{a}_{jW} \right| \right\}$$

$$s.t. \begin{cases} \Sigma_{j=1}^{n} R(\tilde{w}_j) = 1 \\ \\ w_j^L \leq w_j^M \leq w_j^U \\ \\ w_j^L \geq 0 \\ \\ j = (1, 2, ...n) \end{cases} \tag{5.5}$$

In Eq. (5.5), $\tilde{w}_B = (\tilde{w}_B^L, \tilde{w}_B^M, \tilde{w}_B^U)$, $\tilde{w}_j = (\tilde{w}_j^L, \tilde{w}_j^M, \tilde{w}_j^U)$, $\tilde{w}_W = (\tilde{w}_W^L, \tilde{w}_W^M, \tilde{w}_W^U)$, $\tilde{a}_{Bj} = (\tilde{w}_{Bj}^L, \tilde{w}_{Bj}^M, \tilde{w}_{Bj}^U)$, and

$$\tilde{a}_{jW} = (\tilde{w}_{jW}^L, \tilde{w}_{jW}^M, \tilde{w}_{jW}^U)$$

Moreover, $R(\tilde{w}_j)$ is the mean value of $\tilde{w}_j$ and can be computed as given in Eq. (5.6)

$$R(\tilde{w}_j) = \frac{w_j^L + 4 * w_j^M + w_j^U}{6} \tag{5.6}$$

The above constraint problem given in Eq. (5.5) can be converted into non-linear constraint optimisation problem as given in Eq. (5.7).

$$min \ \kappa$$

$$\begin{cases} \left| \frac{\tilde{w}_B}{\tilde{w}_j} - \tilde{a}_{Bj} \right| \leq \kappa \\ \\ \left| \frac{\tilde{w}_j}{\tilde{w}_W} - \tilde{a}_{jW} \right| \leq \kappa \\ \\ \Sigma_{j=1}^{n} R(\tilde{w}_j) = 1 \\ \\ w_j^L \leq w_j^M \leq w_j^U \\ \\ w_j^L \geq 0 \\ \\ j = 1, 2, ...n \end{cases} \tag{5.7}$$

where $\kappa = (\kappa^L, \kappa^M, \kappa^U)$ and $\kappa^L \le \kappa^M \le \kappa^U$,, and it is supposed that $\kappa = (\zeta^*, \zeta^*, \zeta^*)$, $\zeta^* \le \kappa^L$. Therefore, Eq. (5.7) can be transformed into Eq. (5.8):

$$min \ \zeta^*$$

$$s.t. \begin{cases} \left| \dfrac{(w_B^L, w_B^M, w_B^U)}{(w_j^L, w_j^M, w_j^U)} - (a_{Bj}^L, a_{Bj}^M, a_{Bj}^U) \right| \le (\zeta^*, \zeta^*, \zeta^*), \ j = 1, 2...., n \\[2mm] \left| \dfrac{(w_j^L, w_j^M, w_j^U)}{(w_W^L, w_W^M, w_W^U)} - (a_{jW}^L, a_{jW}^M, a_{jW}^U) \right| \le (\zeta^*, \zeta^*, \zeta^*), \ j = 1, 2...., n \\[2mm] \Sigma_{j=1}^n R(\tilde{w}_j) = 1 \\[2mm] w_j^L \le w_j^M \le w_j^U \\[2mm] \tilde{w}_j = (w_j^L, w_j^M, w_j^U) \\[2mm] w_j^L \ge 0 \\[2mm] j = 1, 2, ...n \end{cases}$$

(5.8)

Eq. (5.8) can be transformed into Eq. (5.9), and its solution provides the optimal weight of n criteria $(\tilde{w}_1, \tilde{w}_2, \tilde{w}_3......, \tilde{w}_n)$.

$$min \ \zeta^*$$

$$\begin{cases} \left| \dfrac{(w_B^L)}{(w_j^U)} - a_{Bj}^L \right| \le \zeta^* \\[2mm] \Sigma_{j=1}^n R(\tilde{w}_j) = 1 \\[2mm] w_j^L \le w_j^M \le w_j^U \\[2mm] \tilde{w}_j = (w_j^L, w_j^M, w_j^U) \\[2mm] w_j^L \ge 0 \\[2mm] j = 1, 2, ...n \end{cases}$$

(5.9)

Step 5: Defuzzification of the fuzzy optimal weights of the n criteria $(\tilde{w}_1, \tilde{w}_2, \tilde{w}_3........, \tilde{w}_n)$ into crisp numbers $(w_1, w_1, w_1........, w_n)$ are done with the help of Eq. (5.6).

Step 6: Check the consistency ratio (CR).

Table 5.2 Consistency index (CI) values [39]

Linguistic terms	Equally preferred (EI)	Weakly preferred (WI)	Fairly preferred (FI)	Strongly preferred (SI)	Absolutely preferred (AI)
$\tilde{a}_{BW}$	(1,1,1)	(2/3,1,3/2)	(3/2,2,5/2)	(5/2,3,7/2)	(7/2,4,9/2)
CI	3.00	3.80	5.29	6.69	8.04

Consistency ratio (CR) is a measure of the consistency of pairwise decision-making process [38]. The consistency ratio can be calculated by using Eq. (5.10).

$$CR = \frac{\zeta^*}{CI} \tag{5.10}$$

The consistency index (CI) is computed with the maximum distance vector of best to worst element and presented in Table 5.2.

The lower value of CR is desirable for high consistency. The value more close to the zero shows that the vectors of BO and OW are consistent. A threshold of CR is 0.10; therefore, a lower than the threshold value is required for consistent results [12].

5.4 Application of the Proposed Framework

The proposed green freight transport practices framework is validated with the Indian freight transport firms. The Indian freight industry is growing at a rate of 13.35% CAGR [35]. In India, most of the road traffic is dependent on the finite petroleum resources, and also there is a huge imbalance between the share of transportation mode, which will impose heavy burden on Indian roads, whereas rail, inland waterway, and air modes are still highly underutilised [42]. As per the Environment assessment agency (EAA), level of CO_2 emission in India is 2.47 billion tonnes in the year 2015 [43]. The NTDPC [37] report developed a framework to balance the railroad balance by 50:50 by the year 2032 and promote green freight transport by developing multimodal services. The proposed research framework shed light on the development of sustainable freight transport system through adoption of green freight transport practices.

The industry experts are selected from the multiple companies providing various logistics value-adding services and multimodal freight transport services in India as well as other Asian countries. The expert's selection criteria include the knowledge of the domain, minimum work experience of 10 years, and participation in various environmental sustainability programs. A total of eight experts are agreed for answering the questionnaire. The proposed method gives good consistency with lesser experts [39]. The individual expert has been identified best, worst, best to others (BO), and others to worst (OW) vector on linguistics scale and given in

Table 5.3. The linguistics values are converted into the corresponding TFN and given in Table 5.4. The fuzzy importance weight of each GFTP is computed with the FBWM method explained in Eq. (5.9). The non-linear optimisation model of important weight computation is developed using LINGO optimisation tool, and the corresponding fuzzy weight values are given in Table 5.5. The conversion of fuzzy importance weight of GFTPs is converted into crisp form by using Eq. (5.6) and given in Table 5.6. The importance weight values are prioritised and given in Table 5.6. For illustration purpose, expert 1 fuzzy weight computation, crisp weight, and consistency ratio are given in Tables 5.5 and 5.6. The non-linear optimisation of weight computation gives fuzzy importance weight of GFTP1 as (0.1858, 0.2020, 0.2164). The corresponding crisp value is computed as

$$W_{GFTP_1} = \frac{0.1858 + (4*0.2020) + 0.2164}{6} = 0.2017.$$

The non-linear programming model gives the objective function ζ^* as 0.8074. The consistency ratio is computed as the distance vector between best and worst vector. For expert 1, the distance vector is AI and the corresponding CI value is 8.04 (see Table 5.2). The consistency ratio values are close to 0.1, claimed that the obtained result is consistent. The overall consistency of the decision-making process is computed as 0.093 < 0.1, reported that the decision-making model is consistent.

$$CR = \frac{0.8074}{8.04} = 0.1$$

5.5 Result Discussion

The FBWM method is used to compute the importance weight and priority of the GFTPs. Table 5.6 delineated the following ranking of the GFTPs; GFTP5> GFTP9> GFTP1> GFTP3> GFTP6> GFTP7> GFTP2> GFTP8. The above prioritisation of green practices shows that competitive pressure from other transport firms' green practices is a motivation for adopting green transport practices. Sureeyatanapas et al.'s [25] research highlighted that competitive pressure from other transport firms is a major driver for green practices adoption for logistics service providers. Green pressure from customers and business organisation significantly motivates logistics firms to adopt green freight transport practices, and a similar finding is claimed by Lin and Ho [10]. Incentivitise green freight transport practices (GFTP9) placed second priority on green practices ranking. The government policymakers can propose adequate incentive policies for transport firms using green technologies in their logistics businesses. Government policymakers can sponsor various green training programs such as eco-driving practices, green knowledge sharing, and green training programs. The support from top management for

Table 5.3 Identification of best, worst, best to others, and other to worst vector (linguistic response)

Expert #	Best practice	GFTP1	GFTP2	GFTP3	GFTP4	GFTP5	GFTP6	GFTP7	GFTP8	GFTP9
Best to other (BO) vector from all experts										
1	GFTP1	EI	SI	FI	SI	FI	SI	AI	SI	FI
2	GFTP3	FI	AI	EI	SI	FI	FI	SI	FI	SI
3	GFTP3	SI	SI	EI	AI	SI	FI	FI	AI	SI
4	GFTP9	FI	AI	FI	SI	FI	FI	SI	FI	EI
5	GFTP9	FI	SI	FI	FI	SI	SI	AI	AI	EI
6	GFTP9	SI	AI	SI	FI	SI	WI	SI	SI	EI
7	GFTP6	WI	SI	FI	SI	FI	EI	SI		SI
8	GFTP1	EI	SI	SI	FI	SI	FI	SI	AI	FI

Expert #	Worst practice	GFTP1	GFTP2	GFTP3	GFTP4	GFTP5	GFTP6	GFTP7	GFTP8	GFTP9
Other to worst (OW) vector from all experts										
1	GFTP7	AI	FI	SI	FI	SI	SI	EI	SI	SI
2	GFTP2	SI	EI	AI	SI	FI	SI	FI	SI	FI
3	GFTP4	SI	FI	AI	EI	FI	SI	FI	FI	WI
4	GFTP2	SI	EI	SI	FI	FI	SI	FI	SI	AI
5	GFTP7	FI	WI	SI	SI	SI	FI	EI	FI	AI
6	GFTP2	SI	EI	FI	SI	FI	FI	SI	FI	AI
7	GFTP8	SI	FI	SI	WI	FI	AI	SI	EI	FI
8	GFTP8	AI	SI	SI	FI	FI	SI	FI	EI	FI

Table 5.4 Conversion of linguistic vectors into equivalent TFNs

Expert #	Best practice	GFTP1	GFTP2	GFTP3	GFTP4	GFTP5	GFTP6	GFTP7	GFTP8	GFTP9
Best to other vector from eight experts										
1	GFTP7	(1,1,1)	(5/2,3,7/2)	(3/2,2,5/2)	(5/2,3,7/2)	(3/2,2,5/2)	(5/2,3,7/2)	(7/2,4,9/2)	(5/2,3,7/2)	(3/2,2,5/2)
2	GFTP3	(3/2,2,5/2)	(7/2,4,9/2)	(1,1,1)	(5/2,3,7/2)	(3/2,2,5/2)	(3/2,2,5/2)	(5/2,3,7/2)	(3/2,2,5/2)	(5/2,3,7/2)
3	GFTP3	(5/2,3,7/2)	(5/2,3,7/2)	(1,1,1)	(7/2,4,9/2)	(5/2,3,7/2)	(3/2,2,5/2)	(3/2,2,5/2)	(7/2,4,9/2)	(5/2,3,7/2)
4	GFTP9	(3/2,2,5/2)	(7/2,4,9/2)	(3/2,2,5/2)	(5/2,3,7/2)	(3/2,2,5/2)	(3/2,2,5/2)	(5/2,3,7/2)	(3/2,2,5/2)	(1,1,1)
5	GFTP9	(3/2,2,5/2)	(5/2,3,7/2)	(3/2,2,5/2)	(3/2,2,5/2)	(5/2,3,7/2)	(5/2,3,7/2)	(7/2,4,9/2)	(7/2,4,9/2)	(1,1,1)
6	GFTP9	(5/2,3,7/2)	(7/2,4,9/2)	(5/2,3,7/2)	(3/2,2,5/2)	(5/2,3,7/2)	(2/3,1,3/2)	(5/2,3,7/2)	(5/2,3,7/2)	(1,1,1)
7	GFTP6	(2/3,1,3/2)	(5/2,3,7/2)	(3/2,2,5/2)	(5/2,3,7/2)	(3/2,2,5/2)	(1,1,1)	(5/2,3,7/2)	(7/2,4,9/2)	(5/2,3,7/2)
8	GFTP1	(1,1,1)	(5/2,3,7/2)	(5/2,3,7/2)	(3/2,2,5/2)	(5/2,3,7/2)	(3/2,2,5/2)	(5/2,3,7/2)	(7/2,4,9/2)	(3/2,2,5/2)
Expert #	Worst practice	GFTP1	GFTP2	GFTP3	GFTP4	GFTP5	GFTP6	GFTP7	GFTP8	GFTP9
Other to worst vector from eight experts										
1	GFTP7	(7/2,4,9/2)	(3/2,2,5/2)	(5/2,3,7/2)	(3/2,2,5/2)	(5/2,3,7/2)	(5/2,3,7/2)	(1,1,1)	(5/2,3,7/2)	(5/2,3,7/2)
2	GFTP2	(5/2,3,7/2)	(1,1,1)	(7/2,4,9/2)	(5/2,3,7/2)	(3/2,2,5/2)	(5/2,3,7/2)	(3/2,2,5/2)	(5/2,3,7/2)	(3/2,2,5/2)
3	GFTP4	(5/2,3,7/2)	(3/2,2,5/2)	(7/2,4,9/2)	(1,1,1)	(3/2,2,5/2)	(5/2,3,7/2)	(3/2,2,5/2)	(3/2,2,5/2)	(2/3,1,3/2)
4	GFTP2	(5/2,3,7/2)	(1,1,1)	(5/2,3,7/2)	(3/2,2,5/2)	(3/2,2,5/2)	(5/2,3,7/2)	(3/2,2,5/2)	(5/2,3,7/2)	(7/2,4,9/2)
5	GFTP7	(3/2,2,5/2)	(2/3,1,3/2)	(5/2,3,7/2)	(5/2,3,7/2)	(5/2,3,7/2)	(3/2,2,5/2)	(1,1,1)	(3/2,2,5/2)	(7/2,4,9/2)
6	GFTP2	(5/2,3,7/2)	(1,1,1)	(3/2,2,5/2)	(5/2,3,7/2)	(3/2,2,5/2)	(3/2,2,5/2)	(5/2,3,7/2)	(3/2,2,5/2)	(7/2,4,9/2)
7	GFTP8	(5/2,3,7/2)	(3/2,2,5/2)	(5/2,3,7/2)	(2/3,1,3/2)	(3/2,2,5/2)	(7/2,4,9/2)	(5/2,3,7/2)	(1,1,1)	(3/2,2,5/2)
8	GFTP8	(7/2,4,9/2)	(5/2,3,7/2)	(5/2,3,7/2)	(3/2,2,5/2)	(3/2,2,5/2)	(5/2,3,7/2)	(3/2,2,5/2)	(1,1,1)	(3/2,2,5/2)

Table 5.5 Fuzzy weight computation and CR of expert 1 response

Expert 1	L	M	U	Crisp weight
GFTP1	0.1858	0.202	0.2164	0.2017
GFTP2	0.08	0.092	0.109	0.0928
GFTP3	0.1278	0.1278	0.1278	0.1278
GFTP4	0.08	0.092	0.109	0.0928
GFTP5	0.1278	0.1278	0.1278	0.1278
GFTP6	0.08	0.0921	0.1098	0.0930
GFTP7	0.0407	0.042	0.043	0.0419
GFTP8	0.08	0.092	0.109	0.0928
GFTP9	0.1278	0.1278	0.1278	0.1278
ζ^*			0.8074	
Consistency ratio		0.1		

Note: L,M, and U are the lower, middle, and upper values of triangular fuzzy number

Table 5.6 Aggregation of fuzzy importance weight and priority rank of GFTPs

GFTPs	L	M	U	Weight	Priority rank
GFTP1	0.128	0.140	0.162	0.141	3
GFTP2	0.067	0.078	0.098	0.080	8
GFTP3	0.132	0.141	0.151	0.141	4
GFTP4	0.090	0.099	0.107	0.099	6
GFTP5	0.194	0.197	0.203	0.197	1
GFTP6	0.122	0.133	0.138	0.132	5
GFTP7	0.076	0.086	0.094	0.085	7
GFTP8	0.072	0.078	0.087	0.079	9
GFTP9	0.130	0.146	0.152	0.144	2

green practices (GFTP1) is ranked third in priority ranking. Findings of this work recommended that scalability of green practices is depend on the top management support. The green knowledge management among the employees can be shared with the support of top management. Ojo et al. [44] reported that poor commitment by the top management is a potential barrier of green supply chain management in construction industry, and [25] shows that organisational support is claimed as a critical factor for the adoption of any new policy, and the top management becomes a key component for green initiatives adoption.

Environmental management system certification (GFTP3) obtained the fourth rank. The EMS certification (ISO: 14000) for freight transport organisation and green certification for freight vehicles would improve the green performance of the firm. Sometimes customers will encourage for ISO:14000 certifications for transport firm. The promotion of multimodal services for long hauling the GFTP6 placed on 5th position in the ranking order. The long hauling by rail mode significantly reduces the emission and improves the environmental sustainability of freight transport system [45]. They also recommended that greening freight transport mode is only achieved by shifting road-based transport to intermodal services. Lammgård [28] concluded that the use of intermodal freight transport services for long haul

Table 5.7 Sensitivity analysis with varying importance weight of GFTP5 from 0.1–0.9

GFTPs	Normal value	0.1	0.2	0.3	0.4	0.5	0.6	0.7	0.8	0.9
GFTP1	0.141	0.159	0.141	0.123	0.106	0.088	0.070	0.053	0.035	0.018
GFTP2	0.080	0.090	0.080	0.070	0.060	0.050	0.040	0.030	0.020	0.010
GFTP3	0.141	0.158	0.141	0.123	0.105	0.088	0.070	0.053	0.035	0.018
GFTP4	0.099	0.111	0.098	0.086	0.074	0.061	0.049	0.037	0.025	0.012
GFTP5	0.197	0.100	0.2	0.3	0.4	0.5	0.6	0.7	0.8	0.9
GFTP6	0.132	0.148	0.131	0.115	0.098	0.082	0.066	0.049	0.033	0.016
GFTP7	0.085	0.096	0.085	0.074	0.064	0.053	0.043	0.032	0.021	0.011
GFTP8	0.079	0.088	0.078	0.069	0.059	0.049	0.039	0.029	0.020	0.010
GFTP9	0.144	0.162	0.144	0.126	0.108	0.090	0.072	0.054	0.036	0.018

Table 5.8 Sensitivity analysis based ranking of GFTPs

GFTPs	0.1	0.194	0.2	0.3	0.4	0.5	0.6	0.7	0.8	0.9
GFTP1	2	3	3	3	3	3	3	3	3	3
GFTP2	8	8	8	8	8	8	8	8	8	8
GFTP3	3	4	4	4	4	4	4	4	4	4
GFTP4	5	6	6	6	6	6	6	6	6	6
GFTP5	6	1	1	1	1	1	1	1	11	
GFTP6	4	5	5	5	5	5	5	5	5	5
GFTP7	7	7	7	7	7	7	7	7	7	7
GFTP8	9	9	9	9	9	9	9	9	9	9
GFTP9	1	2	2	2	2	22		2	2	2

is an excellent way to reduce CO_2 emissions and improve the environmental performance. Promoting collaborative green transport practices with other transport organizations places on 6th position. El Baz and Laguir [46] reported that supply chain collaboration on green initiatives improves the environmental performance of logistics service providers. Reduction of empty vehicle run (GFTP7) is placed on the 7th place. Publication of annual sustainability report (GFTP2) and environmental awareness programs (GFTP8) are placed on lower rank in the priority list.

The stability of the above-ranking order of GFTPs is tested through the sensitivity analysis. The sensitivity analysis helps to visualise the changes in the ranking order of GFTPs by varying weight of highest priority GFTP (in this case, GFTP5). The small variation in importance weight of most influential factor changes the corresponding weight of other factors [47]. Table 5.7 delineated the GFTP5 (ranked 1 green practice) weight variation from 0.1 to 0.9 and the corresponding changes in other GFTPs' weights. The ranking order of each sensitivity run is presented in Table 5.8. Sensitivity analysis of green freight transport practices shows that the rank of the GFTPs is the same in most of the runs. The graphical representation of rank variation is presented in Fig. 5.2.

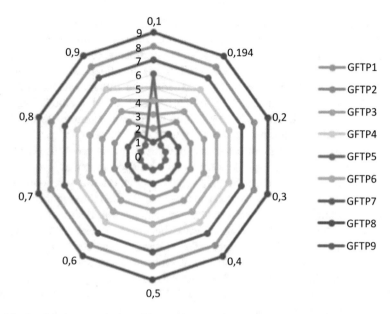

Fig. 5.2 Graphical representation of the ranking variation

5.6   Managerial Implications

This chapter presents the green freight transport practices assessment. Based on the above research findings, the following managerial implications are presented:

- The freight logistics managers can include environmental sustainability practices in their decision-making process. The incentive policies for green freight transport practices attract the LSPs to adopt green practices.
- The presented research framework can help to develop specific policies to improve the green performance of the freight transport industry.
- The obtained result shows that a transport firm can adopt collaborative transport practices for improving sustainability performance.
- The used method can be extended to making a decision support tool for the freight transport industry.

5.7   Conclusion

The green fright transportation system is required to improve the sustainability performance of the freight transport organisations. The freight transport organisations working in emerging economies face challenges to improve their environmental performance due to the lack of unified green transport practices. This chapter tries to shed light on the importance of green freight transport practices (GFTPs) in emerging economies in the freight transport industry. This chapter includes nine GFTPs for assessing the importance of environmental sustainability inclusion in the decision-making process. The GFTPs are identified with the literature review and validated with the industry experts. The importance weight of the GFTPs is determined with the help of the novel MCDM method, i.e., fuzzy best–worst method (FBWM). The application of FBWM required less number of pairwise comparisons and improved consistency ratio. The obtained result reported that competitive pressure from other transport firms' green practices (GFTP5) is important motivation for adopting green freight transport practices. Incentives for green freight transport practices are judged as important criteria for adopting green freight transport practices. The model robustness is checked through the sensitivity analysis and reported that in most of the experiments run, a similar priority has been obtained. Based on the obtained result, this chapter presents the managerial implications for the logistics managers. The presented research framework having certain limitations such as the inclusion of more practices and to measure their interrelationship through other MCDM methods such as analytic network process (ANP) and interpretive structure modelling (ISM). The proposed framework can be tested to other emerging economies for generalization of the findings.

Acknowledgments The authors would like to thank all industry experts for their valuable time and genuine responses. The authors also thank the Editors of this book and the reviewer for giving insightful comments to improve our work. The Ministry of Human Resource Development (MHRD), Government of India, through IIT Roorkee, financially supports this work under a PhD fellowship grant number MHRD/IITR/DoMS/16918015.

References

1. Demir, E., Huang, Y., Scholts, S., Van Woensel, T.: A selected review on the negative externalities of the freight transportation: modeling and pricing. Transp. Res. E Logist. Transp. Rev. **77**, 95–114 (2015)
2. Kumar, A., Anbanandam, R.: Development of social sustainability index for freight transportation system. J. Clean. Prod. **210**, 77–92 (2019)
3. World Commission on Environment and Development: Our Common Future - Report of the World Commission on Environment and Development (The Brundtland Report). Medicine, Conflict and Survival (1987). https://doi.org/10.1080/07488008808408783
4. Ratanavaraha, V., Jomnonkwao, S.: Trends in thailand co2 emissions in the transportation sector and policy mitigation. Transp. Policy **41**, 136–146 (2015)
5. Liu, H., Wu, J., Chu, J.: Environmental efficiency and technological progress of transportation industry-based on large scale data. Technol. Forecast. Soc. Change **144**, 475–482 (2019)
6. Abbasi, M., Nilsson, F.: Developing environmentally sustainable logistics: exploring themes and challenges from a logistics service providers' perspective. Transp. Res. D Transp. Environ. **46**, 273–283 (2016)
7. Centobelli, P., Cerchione, R., Esposito, E.: Developing the wh2 framework for environmental sustainability in logistics service providers: a taxonomy of green initiatives. J. Clean. Prod. **165**, 1063–1077 (2017)

8. Colicchia, C., Marchet, G., Melacini, M., Perotti, S.: Building environmental sustainability: empirical evidence from logistics service providers. J. Cleaner Production **59**, 197–209 (2013)
9. Ellram, L.M., Golicic, S.L.: Adopting environmental transportation practices. Transp. J. **54**(1), 55–88 (2015)
10. Lin, C.-Y., Ho, Y.-H.: Determinants of green practice adoption for logistics companies in China. J. Bus. Ethics **98**(1), 67–83 (2011)
11. Seuring, S., Müller, M.: From a literature review to a conceptual framework for sustainable supply chain management. J. Cleaner Production **16**(15), 1699–1710 (2008)
12. Rezaei, J.: Best-worst multi-criteria decision-making method. Omega **53**, 49–57 (2015)
13. Wolf, C., Seuring, S.: Environmental impacts as buying criteria for third party logistical services. Int. J. Phys. Distrib. Logist. Manag. 40(1–2), 84–102 (2010). https://doi.org/10.1108/09600031011020377
14. Evangelista, P., Colicchia, C., Creazza, A.: Is environmental sustainability a strategic priority for logistics service providers? J. Environ. Manag. **198**, 353–362 (2017)
15. Perotti, S., Zorzini, M., Cagno, E., Micheli, G.J.L.: Green supply chain practices and company performance: the case of 3PLS in Italy. Int. J. Phys. Distrib. Logist. Manag. **42**(7), 640–672 (2012)
16. Björklund, M.: Influence from the business environment on environmental purchasing—drivers and hinders of purchasing green transportation services. J. Purch. Supply Manag. **17**(1), 11–22 (2011)
17. Lieb, K.J., Lieb, R.C.: Environmental sustainability in the third-party logistics (3pl) industry. Int. J. Phys. Distrib. Logist. Manag. **40**(7), 524–533 (2010)
18. Liesen, A., Hoepner, A.G., Patten, D.M., Figge, F.: Does stakeholder pressure influence corporate GHG emissions reporting? Empirical evidence from Europe. Account. Audit. Account. J. **28**(7), 1047–1074 (2015)
19. Affi, M., Derbel, H., Jarboui, B.: Variable neighborhood search algorithm for the green vehicle routing problem. Int. J. Ind. Eng. Comput. **9**(2), 195–204 (2018)
20. Kumar, A., Anbanandam, R.: Multimodal freight transportation strategic network design for sustainable supply chain: an or prospective literature review. Int. J. Syst. Dynam. Appl. **8**(2), 19–35 (2019)
21. Oberhofer, P., Dieplinger, M.: Sustainability in the transport and logistics sector: lacking environmental measures. Bus. Strategy Environ. **23**(4), 236–253 (2014)
22. Ramanathan, U., Bentley, Y., Pang, G.: The role of collaboration in the UK green supply chains: an exploratory study of the perspectives of suppliers, logistics and retailers. J. Clean. Prod. **70**, 231–241 (2014)
23. L.M. Ellram, S.L. Golicic, The role of legitimacy in pursuing environmentally responsible transportation practices. J. Clean. Prod. **139**, 597–611 (2016)
24. Lam, J.S.L., Dai, J.: Environmental sustainability of logistics service provider: an ANP-QFD approach. Int. J. Logist. Manag. **26**(2), 313–333 (2015)
25. Sureeyatanapas, P., Poophiukhok, P., Pathumnakul, S.: Green initiatives for logistics service providers: an investigation of antecedent factors and the contributions to corporate goals. J. Cleaner Production **191**, 1–14 (2018)
26. Eng-Larsson, F., Kohn, C.: Modal shift for greener logistics–the shipper's perspective. Int. J. Phys. Distrib. Logist. Manag. **42**(1), 36–59 (2012)
27. Janic, M.: Modelling the full costs of an intermodal and road freight transport network. Transp. Res. D Transp. Environ. **12**(1), 33–44 (2007)
28. Lammgård, C.: Intermodal train services: a business challenge and a measure for decarbonisation for logistics service providers. Res. Transp. Bus. Manag. **5**, 48–56 (2012)
29. Wong, E.Y.C., Tai, A.H., Zhou, E.: Optimising truckload operations in third-party logistics: a carbon footprint perspective in volatile supply chain. Transp. Res. D Transp. Environ. **63**, 649–661 (2018)
30. Liimatainen, H., Nykänen, L., Arvidsson, N., Hovi, I.B., Jensen, T.C., Østli, V.: Energy efficiency of road freight hauliers—a nordic comparison. Energy Policy **67**, 378–387 (2014)

31. González-Benito, J., González-Benito, Ó.: The role of stakeholder pressure and managerial values in the implementation of environmental logistics practices. Int. J. Product. Res. **44**(7), 1353–1373 (2006)
32. Sarkis, J., Zhu, Q.: Environmental sustainability and production: taking the road less travelled. Int. J. Prod. Res. **56**(1–2), 743–759 (2018)
33. Bask, A., Rajahonka, M.: The role of environmental sustainability in the freight transport mode choice: a systematic literature review with focus on the EU. Int. J. Phys. Distrib. Logist. Manag. **47**(7), 560–602 (2017)
34. Kumar, A., Anbanandam, R.: Assessment of environmental and social sustainability performance of the freight transportation industry: an index-based approach. Transp. Policy (2020). https://doi.org/10.1016/j.tranpol.2020.01.006
35. McKinsey & Company: Building India-Transforming the Nation's Logistics Infrastructure. Chennai: McKinsey & Company (2016). Retrieved 10-03-2018, from https://www.mckinsey.com/~/media/mckinsey/industries/travel%20transport%20and%20logistics/our%20insights/transforming%20indias%20logistics%20infrastructure/building_india%20transforming_the_nations_logistics_infrastructure.ashx
36. Bask, A., Rajahonka, M., Laari, S., Solakivi, T., Töyli, J., Ojala, L.: Environmental sustainability in shipper-lsp relationships. J. Clean. Prod. **172**, 2986–2998 (2018)
37. NTDPC: India Transport Report -Moving India to 2032. Planning Commission. New Delhi: Routledge (2014). Retrieved 10-03-2018, from https://planningcommission.nic.in/reports/genrep/NTDPC_Vol_01.pdf
38. Rezaei, J.: Best-worst multi-criteria decision-making method: some properties and a linear model. Omega **64**, 126–130 (2016)
39. Guo, S., Zhao, H.: Fuzzy best-worst multi-criteria decision-making method and its applications. Knowl.-Based Syst. **121**, 23–31 (2017)
40. Govindan, K., Khodaverdi, R., Jafarian, A.: A fuzzy multi criteria approach for measuring sustainability performance of a supplier based on triple bottom line approach. J. Clean. Prod. **47**, 345–354 (2013)
41. Tseng, M.-L., Lin, Y.-H., Chiu, A.S.F.: Fuzzy AHP-based study of cleaner production implementation in Taiwan PWB manufacturer. J. Clean. Prod. **17**(14), 1249–1256 (2009)
42. Kumar, A., Ramesh, A.: Modelling intermodal freight transportation promotion for sustainable supply chain in India. In: Sustainable Operations in India, pp. 115–137. Springer, Berlin (2018)
43. European Environment Agency (EEA): ENER 016 - Final energy consumption by sector and fuel. Copenhagen: European Environment Agency (2017). Retrieved from https://www.eea.europa.eu/themes/climate/trends-and-projections-in-europe/trends-and-projections-in-europe-2017
44. Ojo, E., Mbowa, C., Akinlabi, E.T.: Barriers in implementing green supply chain management in construction industry. In: International Conference on Industrial Engineering and Operations Management (2014)
45. Eng-Larsson, F., Norrman, A.: Modal shift for greener logistics- exploring the role of the contract. Int. J. Phys. Distrib. Logist. Manag. **44**(10), 721–743 (2014)
46. El Baz, J., Laguir, I.: Third-party logistics providers (TPLs) and environmental sustainability practices in developing countries: the case of Morocco. Int. J. Oper. Prod. Manag. **37**(10), 1451–1474 (2017)
47. Luthra, S., Garg, D., Haleem, A.: Critical success factors of green supply chain management for achieving sustainability in Indian automobile industry. Prod. Plan. Control **26**(5), 339–362 (2015)

Chapter 6
A Novel Hybrid Multi-objective Optimization Approach for Sustainable Delivery Systems with a Case Study of Izmir

Hamdi Giray Resat

Abstract The main novelty of this study is to present a novel two-stage solution method designed for sustainable last-mile delivery systems in urban areas. A proposed hybrid solution methodology includes a multi-criteria decision-making system to select the most efficient logistics providers by considering different performance indicators and a mixed integer linear optimization model in operations of last-mile cargo distributions by drones within metropolitan areas by considering time windows for customer services. We present the multi-objective modelling approach, data analysis and outline important characteristics of the mathematical programming problem to minimize transportation cost (in the meantime, carbon dioxide emissions) and total sustainability score of the system by using epsilon-constraint method to find out the Pareto frontier. The proposed solution methodology is applied to an illustrative case by using real-life data of one of the metropolitans in Turkey. The approach is shown as comparative analysis, after defining some preprocessing, symmetry breaking steps, valid inequalities and logic cuts.

6.1 Introduction

Companies operating in last-mile deliveries search for new and adaptive systems to survive under harsh competitive conditions. Adaptation for the last-minute changes or maintenance/breakdown cases leads to excess expenditures, longer delivery duration and increased rate of unsatisfied customers. Therefore, proposed systems should ensure redesign and preparation of advanced and dynamic plans for routing problems because customers expect rapid interactions with their suppliers in delivery or return processes. This situation leads to increase loyalty and satisfaction

H. G. Resat (✉)
Department of Industrial Engineering, Izmir University of Economics, Izmir, Turkey
e-mail: giray.resat@ieu.edu.tr

© Springer Nature Switzerland AG 2020
H. Derbel et al. (eds.), *Modeling and Optimization in Green Logistics*,
https://doi.org/10.1007/978-3-030-45308-4_6

ratios. Apart from the individual effect of last-mile delivery operations over citizens, there is a significant effect over energy consumption and environmental issues (such as greenhouse gas emissions, carbon footprint, etc.). The efficiency of the last-mile operations not only influences the profitability of retailing but also affects environmental and social performance criteria such as emissions and traffic congestion in regions. For instance, traffic caused by urban cargo delivery accounts for about 10–15% of kilometres travelled in city centres and emits approximately 6% of all transport-related GHG emissions [1]. 20–25% of freight vehicle kilometres is related to goods leaving urban areas, and 40–50% is related to incoming goods. The remaining percentage relates to internal exchange (i.e., goods having both their origin and their destination within the city) [2].

Developed models or systems should also consider the effect of energy usage and pollution reductions in delivery processes under certain scenarios. One of the innovative solutions for this purpose is the usage of drones or unmanned aerial vehicles in last-mile delivery processes. This concept is getting more attention and popularity after observing some real-life cases (such as Amazon Inc., DHL, etc.). Although there are many advantages of this newly developed system over management of cost, time and carbon emissions, some critical points occur when real-life applications are considered, for example, storage problems of drones at distribution or transshipment centres, bad weather conditions, personal/private data protection, range and capacity limitations of drones, etc. However, when comparing heavy/medium trucks in city centres or at the location of final customers in delivery systems, drones offer significant operational advantages. A drone system can deliver goods directly to the person who ordered it or pick up the order for cancellation cases in less than 30 min, and this duration reduces lead time significantly and eliminates route restrictions and many other logistics obstacles. Distribution of cargo in a combination of both trucks and drones simultaneously is a very complex problem and needs very extensive studies in terms of solution methodologies. Although vehicle routing problem considering different perspectives (such as number of vehicles, variety of vehicles, capacity of the vehicles, time windows of customers, type of service provided, number of warehouses, uncertainty in distribution times) has been studied extensively, inclusion of drones in delivery process is a very hot topic and there is still serious gap in literature in terms of modelling and computational side.

6.2 Literature Review

The survey of Lin et al. [3] gives and analyses an exhaustive literature review over the vehicle routing problems and classifies them into different application domains. According to this survey, green vehicle routing problems should concern energy usage of the systems to reduce fuel consumption that directly affects the greenhouse gas emissions and increases transportation efficiency. In this perspective, Bektas and Laporte [4] present an extended version of classical VRP problem by considering the amount of greenhouse emissions, travel times and total transportation costs. In their study, they developed a mathematical model with comprehensive objective functions to indicate the vehicle loads and speeds. In addition, Conrad and Figliozzi [5] present a routing problem for electric vehicles to decide on the locations of charging stations as the Recharging Vehicle Routing Problem (RVRP). They try to decide on recharging operations of the electric vehicles during the transportation operations in which vehicles can be recharged in the locations of customers instead of dedicated stations. Therefore, there can be some savings in the total transportation times because there can be both loading activities and recharging of the vehicles instead of spending time at the dedicated stations for recharging. Erdogan and Miller-Hooks [6] introduce the Green Vehicle Routing Problem (G-VRP) by considering the recharging of the vehicles with limited fuel capacities. In G-VRP problem, vehicles are eliminated from the risk of running out of fuel and service time of each customer and maximum duration restriction is posed on each route. After this study, Schneider et al. [7] propose an updated version of G-VRP by including customer time windows for the delivery operations in which electric vehicles are used. Murray and Chu [8] introduce the vehicle routing problem with drones (VRPD) in which a drone collaborates with a truck to distribute customer parcels within minimum delivery time. A mixed integer linear programming (MILP) formulation and a heuristic adopting "Truck First, Drone Second" idea are proposed and tested on the instances with 10 customers. Ponza [9] presents a solution methodology for VRPD problems by using both MILP model and simulated annealing algorithm. Ferrandez et al. [10] propose an optimization model of a truck–drone system in bicycle delivery networks by using the K-means algorithms to find the most efficient launch locations as well as using a genetic algorithm to assign the truck route between starting nodes. Wang et al. [11] introduce a more general problem considering multiple trucks and drones with the objective of minimizing the total duration of the delivery mission. Carlsson and Song [12] enhance the VRPD by relaxing the restrictions over meeting points of the unmanned aerial vehicles, such as drones can provide services either at final destination nodes or over trucks on their routes. Dorling et al. [13] propose Drone Delivery Problem (DDP) by considering two VRP-based models that are Minimum-Time-DDP (MT-DDP) and Minimum-Cost-DDP (MC-DDP). Agatz et al. [14] study the VRPD with the objective of minimizing total logistics cost by using route-first, cluster-second heuristics based on local search and dynamic programming. Wang and Sheu [15] propose an MILP model and a branch-and-price algorithm for VRPD. Different from the classical VRP, there are two types of vehicles in proposed model: a drone may have multiple times of flying and landing, each of which may be associated with a different truck and a truck may launch and collect multiple drones at different times and locations. Karak and Abdelghany [16] develop a methodology that extends the classic Clarke and Wright algorithm to solve the hybrid vehicle–drone routing problem. Kitjacharoenchai et al. [17] present a MILP formulation with the objective of minimizing the arrival time of both trucks and drones at the depot after completing the deliveries. A new heuristics algorithm is also developed to solve large-sized problems containing up to a hundred locations.

Considering a single objective function in getting a feasible solution set for such complicated problems may not be so realistic in real-life cases. If harsh competition conditions are considered, objective of minimizing total operational cost gets much more importance, but inclusion of sustainability factors (such as time, GHG emissions, customer satisfaction rates, etc.) will lead to more realistic models and solution sets for decision makers. Therefore, inclusion of other objective functions will convert the problem into bi- or multi-objective problems. There are several methodologies to get non-dominated solution sets or Pareto frontiers for multi-objective problems, which are weighted sum and epsilon-constraint models in general. In this study, the epsilon-constraint method is used because the weighting method may sometimes miss some of the efficient solutions in some feasible regions [18]. In the therefore, a set of solutions can be found by improving one objective and worsening others. Several approaches are proposed in the literature for the epsilon-constraint method. The augmented epsilon-constraint (AUGMECON) method by Mavrotas [18] and its improved version (AUGMECON2) by Mavrotas and Florios [19] propose solution methods to find a set of exact solutions in multi-objective problems by adding slack variables into the objective functions that are taken as constraints and penalty term into the main objective function. Fattahi and Turkay [20] present a novel one-direction search (ODS) method to find an exact non-dominated frontier of bi-objective MILPs.

If sustainability factors are considered, multi-criteria decision-making models should be used for pairwise comparisons that can allow decision makers to weight coefficients and compare alternatives with relative ease [21]. Noorizadeh [22] proposes data envelopment analysis (DEA) to select green suppliers to reduce carbon emissions. According to Kannan et al. [23], goal programming (GP) is a useful method to solve multi-criteria decision-making problems since it can easily be applied to solve complex problems. Jolai et al. [24] obtain the efficiency ranks of suppliers and choose the higher ranked ones by using a fuzzy multi-criteria method. Then, they calculate order quantities of suppliers using a multi-objective mathematical model. Ku et al. [25] present an approach combined fuzzy analytic hierarchy process (FAHP) with fuzzy goal programming (FGP) methods for solving supplier selection problem. Boran et al. [26] use a hybrid approach combined intuitionistic fuzzy sets with Technique for Order Preference by Similarity to Ideal Solution (TOPSIS) to apply a decision-making problem in supplier selection. Liu and Zhang [27] propose a new method that is a combination of entropy weight and an improved Elimination and Choice Translating Reality (ELECTRE) III method. Weights of each indicator are determined based on entropy, and all alternative suppliers are ranked according to their strong abilities. Koksalan and Ozpeynirci [28] propose an interactive approach combined the UTADIS with approach of Koksalan and Ulu [29] to sort non-reference and reference alternatives. UTADIS is used for estimating the additive utility function that uses alternatives assigned to categories. To prioritize supply chain risks, Prasanna and Kumanan [30] propose an approach combining AHP with Preference Ranking Organization Method for Enrichment Evaluation (PROMETHEE) for plastic industry.

The main novelties of this study are to

- design a multi-objective optimization framework in order to provide alternative solution sets with different possible routes and schedules;
- develop an MILP model for drone-based routing problem with time windows for last-mile delivery systems;
- consider sustainability concept by integrating as much as possible performance indicators of logistics providers into proposed model by using TOPSIS method;
- implement a hybrid methodology into the proposed model and validate it by using real-life data sets.

The rest of the paper is structured as follows: after this comprehensive literature review and problem definition section, the details of proposed methodology for multi-objective hybrid model will be given in Sect. 6.3. Computational details and data used in the model are given in Sect. 6.4. The details of multi-criteria decision analysis model and development of the MILP model are shared in Sect. 6.5. After obtaining outputs of study for decision makers, some future works and assessment of the findings will be given in Sect. 6.6.

6.3   Methodology

A two-step hybrid approach is proposed in this study to be able to add the largest number of indicators that can be used in the mathematical model of sustainable last-mile delivery processes. The main steps of the methodology are shared in Fig. 6.1.

The proposed methodology includes mainly three steps: first one is the collection of relevant data for logistics provider selection in last-mile delivery systems from final customers. The main critical points in data collection part are to assess criteria for logistics provider selection of final customers in cargo deliveries. Key performance indicators of the delivery operations affecting customer decisions and used in the proposed model are indicated in Table 6.1.

The defined indicators given in Table 6.1 are obtained by using semi-structured interview method. The surveys are conducted in both verbal and written forms in a group of 250 customers and collected scores of different cargo delivery companies for different performance indicators. Final customers filled up the surveys and questionnaires in a range of 1–10. Score 1 indicates the decision of "*not relevant*" and Score 10 indicates "*major*" impact in supplier selection processes. After collecting scores for different logistics providers, final scores of different performance indicators in supplier selection are taken as an average of all scores obtained from all surveys under the same criteria. This means that let us assume that set of α indicates the set of performance indicators, set of k shows the set of logistics providers and i indicates the set of final customers. Therefore, parameter of $\Psi_{k\alpha i}$ shows the score of logistics provider k for performance indicator α given

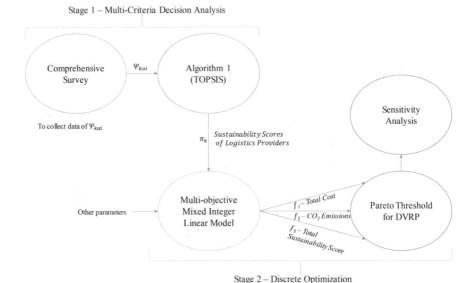

Fig. 6.1 Structure of the proposed methodology

Table 6.1 List of key performance indicators of last-mile delivery operations in customer decisions

Criteria 1	Unit price per kg cargo
Criteria 2	Compliance with time windows
Criteria 3	Lead times of the deliveries
Criteria 4	Traceability of cargo
Criteria 5	Safety of cargo
Criteria 6	Accessibility to the company via mobile/web channels
Criteria 7	Air pollution policies
Criteria 8	Noise pollution policies
Criteria 9	Technological infrastructure
Criteria 10	Proximity of regional office
Criteria 11	Availability of pick-up services
Criteria 12	Standardized services
Criteria 13	Geographical range for delivery
Criteria 14	Policies over security of private data

by customer i. $\frac{\sum_{i \in I} \Psi_{kai}}{|I|}$ gives the score table of each provider per performance indicator, and this table will be the main input of the TOPSIS system.

In the second step of proposed model, a complex decision problem is structured at hierarchical levels and decision alternatives are generated to decrease complexity in multi-criteria problems by using TOPSIS. Decision makers predefine associated weights, and alternative solution sets are performed by using different weights.

Algorithm 1 given below gets final scores of different last-mile delivery companies under defined performance indicators [31].

Algorithm 1:

0: Identification of alternative logistics providers (k) and performance indicators (α)

1: Compute $N_{k\alpha} = \left\{ \frac{\sum_{i \in I} \Psi_{k\alpha i}}{|I|} \right\}^2$, $\qquad \forall k \in K, \; \forall \alpha \in A$

2: Compute $N_{k\alpha}^* = \frac{N_{k\alpha}}{\sqrt[2]{\sum_{\alpha \in A} N_{k\alpha}}}$, $\qquad \forall k \in K, \; \forall \alpha \in A$

3: Calculate variance of weights $V_\alpha = \frac{1}{|A|} \sum_{k \in K} \left(N_{k\alpha}^* - N_{k\alpha} \right)^2$, $\; \forall \alpha \in A$

4: **for** $\alpha \in A$ **do**

5: Obtain weights $\chi_\alpha = \frac{V_\alpha}{\sum_{\alpha \in A} V_\alpha}$, such that $\sum_{\alpha \in A} \chi_\alpha = 1$, $\; \forall \alpha \in A$

6: Compute $\chi_\alpha . N_{k\alpha}^*$, $\forall k \in K$, $\forall \alpha \in A$

7: Determine $max \left\{ \chi_\alpha . N_{k\alpha}^* \right\}$ and $min \left\{ \chi_\alpha . N_{k\alpha}^* \right\}$, $\; \forall \alpha \in A$

8: Compute $\pi_k = \frac{\sum_{\alpha \in A} \left\{ \chi_\alpha . N_{k\alpha}^* - min \{ \chi_\alpha . N_{k\alpha}^* \} \right\}^2}{\sum_{\alpha \in A} \left\{ \chi_\alpha . N_{k\alpha}^* - min \{ \chi_\alpha . N_{k\alpha}^* \} \right\}^2 + \sum_{\alpha \in A} \left\{ \chi_\alpha . N_{k\alpha}^* - max \{ \chi_\alpha . N_{k\alpha}^* \} \right\}^2}$, $\; \forall k \in K$

After collecting final performance scores of different companies (π_k), final scores of companies are given into mathematical model as parameter. Pareto frontier for sustainable distribution channels in city centres is tried to be obtained.

As indicated in the study of Resat [32], green zones should be created within the city centres in order to achieve more environmental cases. In this study, major assumption is that only drone deliveries are allowed to operate in these regions and that carbon emissions caused by high transportation activities will decrease in these highly crowded and carbon-dense areas. Logistics operations start from the distribution centres (*DCs*) located outside of the city centres, and single type of product (a standard parcel) is considered. It is only allowed that a single type of medium truck from distribution centres to transfer points carries the cargo. The capacities of these trucks are considered as constant and their capacities are just a ton. When trucks come to transfer points (*TPs*), there will be transfer from conventional vehicles to the drones. The capacities of drones are also assumed as constant. The degree of extra warehousing requirement for drone delivery dominates the comparison of drone and truck-based scenarios, because non-operating drones should also be stored in some specific warehouses. For servicing an urban area with on-demand delivery, two main approaches have been proposed. The first is to locate distribution centres such that all of the service area is within delivery range of a distribution centre. The second is to establish way stations such that drones can fly from one to another and exchange batteries in a series of hops from distribution centre to customer destination [33]. In this study, way stations (transfer points) are considered and drones are stored at these places, and delivery operations start from these stations. After the transfer points, last-mile deliveries are made by using only drones. However, due to lower the battery lives of drones and flight ranges, there will be some options to recharge these drones during the routing. In our problem, it is assumed that each drone can be recharged at only recharging stations. In addition, there is no limitation in the recharging ratios that means that drones can be recharged less than the maximum level. Drone batteries are recharged with constant ratios, and each battery has a constant capacity.

Assumptions
- All of the customers have to be visited once during a day;
- All the routes are started in the transfer points and ended again in the transfer points (not exactly the same one);
- Different logistics providers can use and store their drones at different transfer points;
- The total demands of the customers in the green zones should not exceed the truck capacities (back order option is eliminated);
- The deliveries should be managed within time windows of the customers;
- Transportation times are directly depending on Euclidean distances between nodes;
- The distances between the routes are taken as symmetric ($d_{ij} = d_{ji}$);
- Demands of the customers are known beforehand.

Model Formulation
The main purpose of this study is to create a multi-objective MILP model to find all possible tours covering all customers exactly once while minimizing total transportation distance (cost), as well as minimize total carbon emissions and maximize total supplier score in terms of sustainability.

Indices
i, j Customers $(i, j = 1, \dots , I)$
n Transfer points $(n = 1, \dots , N)$
m Distribution centres $(m = 1, \dots , M)$
k Logistic providers $(k = 1, \dots , K)$
v Drones $(v = 1, \dots , V)$

Scalars
Cap^{elec} Maximum load capacity of drones
Cap^{conv} Maximum load capacity of conventional trucks
γ Constant charge consumption rate
β Constant recharge rate
max Maximum number of trucks used outside of the green zones
Q Maximum battery capacity
$trans$ Fixed transfer cost in the transfer points
Inv Inventory cost keeping a drone at transfer points
π The pi constant
ζ Overall power efficiency of the drone
n Number of rotors
ρ The density of air
g The gravitational constant
A The projected area of component
D The diameter of rotors

m_{body}	The masses of the drone body
m_{bat}	The masses of the drone battery
m_{cargo}	The masses of package carried by drone
v_a	Average speed of air
c_D	Drag coefficient

Parameters

d'_{ij}	Distance between customer i and j within green zones
d''_{mn}	Distance between distribution centre m and transfer point n outside of green zones
c''_{mn}	Unit transportation cost from distribution centre m to the transfer point n
c'_{ij}	Unit transportation cost between nodes i and j within green zones
max_{kn}	Maximum number of drones per supplier k departing from the transfer point n
FC_n	Fixed cost of opening transfer point n
S^v	Speed of the drone v
q_i	Demand of node i
e_i	Opening time of node i
l_i	Closing time of node i
s_i	Service duration of customer i
π_k	Overall sustainability coefficient of supplier k

Variables

$x_{mn} \in R^+$	Carried quantity from distribution centre m to transfer point n
$h^v_{ik} \in R^+$	Arrival time of the drone v *of* supplier k to customer i
$z^v_{ik} \in R^+$	Remaining charge level of the drone v *of* supplier k on arrival to customer i

$$w^v_{ijk} = \{ \begin{matrix} 1 & \text{if an arc between nodes } i \text{ and } j \text{ is travelled by drone } v \text{ of supplier } k \\ 0 & \text{otherwise} \end{matrix}$$

$$y_n = \{ \begin{matrix} 1 & \text{if transfer point } n \text{ is opened} \\ 0 & \text{otherwise} \end{matrix}$$

$$\theta^v_{ink} = \{ \begin{matrix} 1 & \text{if customer } i \text{ is served by transfer point } n \text{ via drone } v \text{ of supplier } k \\ 0 & \text{otherwise} \end{matrix}$$

$$Min \; f_1 = \sum_{i\in I} \sum_{j\in I, i\neq j} \left\{ \sum_{k\in K} \sum_{v\in V} d'_{ij} w^v_{ijk} c'_{ij} \right\} + \sum_{m\in D} \sum_{n\in N} \left\{ x_{mn} d''_{mn} c''_{mn} \right\}$$

$$+ \; trans \left\{ \sum_{m\in D} \sum_{n\in N} x_{mn} \right\} + \sum_{n\in N} \left\{ y_n FC_n \right\}$$

$$+ \sum_{k\in K} \sum_{n\in N} \left\{ max_{kn} - \sum_{i\in I} \sum_{v\in V} \theta^v_{ink} \right\} Inv$$

$$\tag{6.1}$$

$$Max \; f_2 = \sum_{i\in I} \sum_{j\in J} \sum_{k\in K} \sum_{v\in V} w^v_{ijk} \pi_k \tag{6.2}$$

$$Min \; f_3 = \sum_{i\in I} \sum_{j\in J} \sum_{k\in K} \sum_{v\in V} \frac{\left\{ \left(m_{body} + m_{bat} + m_{cargo} \right) g + \frac{1}{2} \rho v_a c_D A \right\}^{3/2}}{\sqrt{\frac{1}{2}\pi \; D^2 \rho \zeta \, S^v}} d'_{ij} w^v_{ijk}$$

$$\tag{6.3}$$

Subject to

$$\sum_{j\in I: i\neq j} \sum_{k\in K} \sum_{v\in V} w^v_{ijk} = 1, \quad \forall i \in I \tag{6.4}$$

$$\sum_{i\in I: j\neq i} \sum_{k\in K} w^v_{ijk} - \sum_{i\in I: j\neq i} \sum_{k\in K} w^v_{jik} = 0, \quad \forall j \in I, \forall v \in V \tag{6.5}$$

$$\sum_{j\in I} \sum_{v\in V} w^v_{njk} \leq max_{kn}, \quad \forall n \in N, \forall k \in K \tag{6.6}$$

$$\sum_{j\in I} \sum_{v\in V} w^v_{jnk} \leq max_{kn}, \quad \forall n \in N, \forall k \in K \tag{6.7}$$

$$0 \leq h^v_{ik} + \left(\frac{d'_{ij}}{S^v} + s_i \right) w^v_{ijk} + \beta \left(Q - z^v_{ik} \right) \leq h^v_{jk}, \quad \forall i, j \in I, \; \forall v \in V, \; \forall k \in K, \; i \neq j$$

$$\tag{6.8}$$

$$e_i \leq h^v_{ik} \leq l_i, \quad \forall i \in I, \forall v \in V, \forall k \in K \tag{6.9}$$

$$0 \le z_{jk}^v \le \left\{ z_{ik}^v - \left(\gamma d_{ij}' \right) \right\} w_{ijk}^v + Q \left(1 - w_{ijk}^v \right), \quad \forall i, j \in I, \ \forall v \in V, \ \forall k \in K, \ i \ne j \tag{6.10}$$

$$z_{nk}^v = Q, \quad \forall n \in N, \quad \forall k \in K, \ \forall v \in V \tag{6.11}$$

$$\sum_{m \in M} \sum_{n \in N} x_{mn} \ge maxCap^{conv} y_n, \quad \forall n \in N \tag{6.12}$$

$$\sum_{i \in I} \sum_{n \in N} \sum_{k \in K} \sum_{v \in V} w_{nik}^v Cap^{elec} \ge \sum_{i \in I} q_i \tag{6.13}$$

$$\sum_{j \in I} \sum_{v \in V} w_{njk}^v \le |I|, \quad \forall k \in K, \forall n \in N \tag{6.14}$$

$$\sum_{k \in K} \sum_{v \in V} \sum_{i \in I} \theta_{nik}^v q_i \le \sum_{m \in M} x_{mn}, \quad \forall n \in N \tag{6.15}$$

$$-\theta_{ink}^v + \sum_{j \in I: j \ne i} \left(w_{ijk}^v + w_{jnk}^v \right) \le 1, \quad \forall i \in I, \ \forall n \in N, \ \forall v \in V, \forall k \in K \tag{6.16}$$

$$w_{ijk}^v \in \{0, 1\}, \quad \forall i, j \in I, \ \forall v \in V, \ \forall k \in K \tag{6.17}$$

$$y_n \in \{0, 1\}, \quad \forall n \in N \tag{6.18}$$

$$\theta_{ink}^v \in \{0, 1\}, \quad \forall i \in I, \ \forall n \in N, \ \forall v \in V, \forall k \in K \tag{6.19}$$

$$x_{mn} \ge 0, \quad \forall m \in M, \ \forall n \in N \tag{6.20}$$

The objective function (6.1) tries to minimize total cost of the last-mile delivery operation and combines five parts. First and second parts show total transportation cost occurred inside and outside of the green zones, respectively. These expressions use total distance travelled inside/outside of green zones in determined time duration. Third part ensures the total transfer cost for transfer of cargo from conventional vehicles to drones, and forth part is about total cost of opening transfer points on the borders of green zones. Last part gives the total inventory holding cost of unused drones kept in transfer points. Objective function (6.2) indicates the second objective function and tries to maximize total sustainability score of the system by using individual sustainability scores of the logistics providers. Objective function (6.3) shows the third objective function by minimizing carbon emissions of the drone delivery system in green zones. Constraint (6.4) ensures that each vertex between customers will be visited exactly once by each drone, so this constraint enforces the drones to complete their tours. Constraint (6.5) ensures that incoming and outgoing flows have to be equal to guarantee the elimination of sub-tours for each drone. Constraint (6.6) indicates that there are at most "*max*" drones going out from transfer points, and Constraint (6.7) shows at most "*max*" drones can come into the transfer points. Constraint (6.8) shows the time balance of arrival of the drones to the nodes. The difference between arrival times of the nodes has to be more than the summation of total travel time and service time within the node for each drone. Constraint (6.9) guarantees that leaving time of nodes should be within the customer time windows (the opening and closing time of nodes). Constraint (6.10) ensures the balance of total energy consumption of the drones while travelling between the nodes and total energy load of the drones should be until the maximum level of the battery. Constraint (6.11) ensures that drones of each logistics provider start with full loaded batteries initially at transfer points. Constraint (6.12) shows the distribution of the cargo from distribution centres to the transfer points, and this constraint forces the model to give a decision about opening transfer points. Constraint (6.13) ensures that total carried cargo has to fulfil the total demand of the customers in the green zone. Constraint (6.14) indicates that total number of visits of supplier k from transfer point n cannot exceed the total number of customers. Constraint (6.15) ensures the fulfilment of the demand of the customers by distributions centres via transfer points. Constraint (6.16) specifies that the customer can be assigned to transfer point only if there is an arc connecting the transfer point and customer. Constraints (6.17)–(6.20) are the variables' domain.

6.4 Computational Experiments

An illustrative example is carried out by using real-life data obtained from final customers. The aim of this case study is to represent two-stage hybrid optimization approach and demonstrate detailed calculations for sustainable delivery systems. While designing the last-mile delivery system, not only cost minimization but also environmental and sustainability effects are considered.

6.4.1 Data

In this section, all necessary data for comprehensive analysis of design and development of proposed sustainable delivery system in the illustrative example are shared.

First, locations of the customers are obtained from web-based mapping service of one of the biggest search engine providers (Google Maps Platform) and given in Table 6.2. The real coordinates of the customers are converted into some integer values, and x- and y-axes of the graph are located at the nodes where their coordination levels are set at lowest (y-axis) and on right-hand side (x-axis) of the sample network. Euclidean distances between distribution centres, transfer points and customers are calculated by using Constraint (6.21).

$$d'_{ij} = \sqrt{\left(x_i^{coor} - x_j^{coor}\right)^2 + \left(y_i^{coor} - y_j^{coor}\right)^2}, \quad \forall i, j \in I \qquad (6.21)$$

Second, the energy usage and environmental impacts of drone delivery system depend substantially on the range of drones and manner of implementation. Some specific parameters used for drone services in this study are listed as follows:

- Standard type of octocopter drone with lithium-based batteries is used in case studies to carry a single parcel from transfer points to final destination of customers and then return empty to the transfer points;
- Cargo capacities carried by standard drone systems can range between 0.3 and 5 kg, and their speed range is between 20 and 120 km/h [34]. Therefore, it is assumed that each drone can carry 5 kg cargo with a speed of 30 km/h in delivery processes. Range of each drone with full battery is assumed as around 2.5 miles (4 km);
- Drones are expected to fly up to max. 120 min with full battery capacity and not carry any payload [33], and therefore it is assumed that their flight capacity can be max. 100 min in the air;
- In the third objective function of the mathematical model related with minimization of greenhouse gas emissions and energy usage in drone-based delivery systems, energy consumption of drones is included into the model. The masses of the drone body, battery and package, the gravitational constant and the total drag force are used in the calculations, and parameters are taken from the study of Stolaroff et al. [33].

Third, some operational limitations and scalars are included in the model, such as number of available drones per logistics provider, warehouse capacities for drones, time windows for each customer, charge consumption and recharge rates, some cost parameters for direct delivery and transshipment operations.

The proposed MILP model for this problem is written in GAMS modelling environment and solved with IBM ILOG CPLEX 12.1 [35]. Both models are executed on a computer with Intel Core I5 2520 M CPU with 2.50 GHz dual

Table 6.2 Coordinates of the final customers

Customer ID	x-Coordinate	y-Coordinate	Customer ID	x-Coordinate	y-Coordinate	Customer ID	x-Coordinate	y-Coordinate
0	4.60	6.35	26	3.90	2.50	51	5.40	5.00
1	3.85	12.10	27	3.00	2.00	52	5.60	4.25
2	3.95	16.60	28	2.80	5.40	53	5.35	4.25
3	3.40	12.30	29	2.60	4.60	54	5.65	3.00
4	2.75	13.00	30	2.60	2.50	55	5.10	4.00
5	5.45	12.45	31	2.65	5.20	56	5.15	1.00
6	4.15	17.50	32	2.30	5.70	57	5.15	3.50
7	4.65	14.50	33	1.50	5.30	58	4.75	3.90
8	4.40	12.35	34	4.25	5.50	59	4.80	3.50
9	3.90	12.40	35	4.10	5.90	60	4.70	5.30
10	3.85	12.50	36	3.95	4.75	61	4.95	5.80
11	3.30	15.00	37	3.50	5.10	62	5.10	6.00
12	2.85	16.50	38	3.70	4.50	63	5.15	5.60
13	2.50	16.00	39	3.65	3.75	64	5.80	4.00
14	2.10	18.00	40	2.90	4.50	65	5.35	5.70
15	2.75	16.90	41	2.65	3.75	66	5.35	1.25
16	3.25	16.75	42	2.00	4.00	67	5.60	6.20
17	3.80	15.30	43	2.40	3.60	68	5.35	6.10
18	5.25	12.00	44	1.00	3.60	69	4.45	3.50
19	4.60	13.00	45	2.45	1.50	70	5.25	3.25
20	4.85	13.00	46	2.10	6.10	71	5.50	5.50
21	5.15	14.00	47	2.75	6.30	72	5.75	5.00
22	5.15	12.25	48	2.85	6.10	73	4.50	4.50
23	4.30	11.50	49	2.85	5.90	74	4.60	2.00
24	5.00	11.50	50	3.40	6.20	75	5.75	2.25
25	5.35	11.60						

core processor and with 4.00 GB of RAM. An optimality gap of 1% is set for the solutions.

6.5   Results

In this section, the details of the illustrative example are given, and main concepts and analysis of solution on an illustrative case are discussed. Case study is started to find final sustainability scores of four different logistics providers (L1–L2–L3–L4) operating in the district of Izmir, Turkey. The data collection methodology is applied more than 250 final customers who take cargo delivery services from these four suppliers. Therefore, the final score table of each provider per performance indicator ($\frac{\sum_{i \in I} \psi_{kai}}{|I|}$) is given in Table 6.3. The scores and associated weights for defined indicators given in Table 6.3 are obtained by using semi-structured interview method for different logistics providers. The most important indicator of logistics provider selection process in last-mile deliveries is the delivery duration of the providers after orders are set by the customers. The safety of cargoes and private data follows this criterion. Least important ones are related with environmental conditions such as different types of pollution.

After final score table is created for logistics providers, data set is given to Algorithm 1 and final scores of the suppliers (given in Table 6.4) are obtained for the mathematical model.

Graphical representation of outputs of the proposed mathematical model is given in Fig. 6.2. In the illustrative case, four distribution centres are defined at the outside of the city centres. Medium trucks are used in delivery of total cargo between distribution centres and transfer points (Node 0–Node 7–Node 46–Node 64). After cargo is arrived to transfer points, individual delivery of the parcels is managed by using drones.

The details of solutions for illustrative example are given in Table 6.5. The details of routes of individual drones of different logistics providers and their corresponding travelled distances are shared. The critical point is that the distances in bold indicate that drones are charged at those routes by using charging stations.

In Fig. 6.3, the details of Pareto frontier of three objective functions are given. Hundred different solution sets are obtained from proposed model for decision makers, and different strategies can be followed to satisfy requirements of the customers as given in Table 6.6.

Table 6.6 includes four different scenarios. First, feasible solution sets are shared if the problem considered as single objective and corresponding values are obtained for different objective functions separately. Then, two different Pareto solution sets are chosen randomly, and their comparison are given. If any stakeholder would like to spend much more money (from 1.1 to 3.7 mil. Euro), they can reduce their carbon emissions by 80.2% and increase total sustainability score of the system by 97.5%.

Table 6.3 Average scores of each provider per each performance indicator

Weights	0.05	0.09	0.20	0.10	0.10	0.05	0.03
	Unit price per kg	Time windows	Lead times	Traceability	Safety of cargos	Accessibility	Air pollution
L1	1	5	8	10	9	2	6
L2	5	6	10	5	9	3	3
L3	9	9	1	8	5	5	7
L4	5	4	4	1	9	7	7
Weights	0.01	0.05	0.05	0.08	0.04	0.05	0.10
	Noise pollution	Technologic infrastructure	Proximity centre	Pick-up services	Standardize services	Geographical range for delivery	Security of private data
L1	10	3	8	2	4	7	7
L2	2	5	7	2	1	2	9
L3	9	3	5	1	1	4	5
L4	7	5	9	1	6	1	3

Table 6.4 Final sustainability scores of the logistics providers

	Total sustainability score
L1	0.61
L2	0.67
L3	0.32
L4	0.26

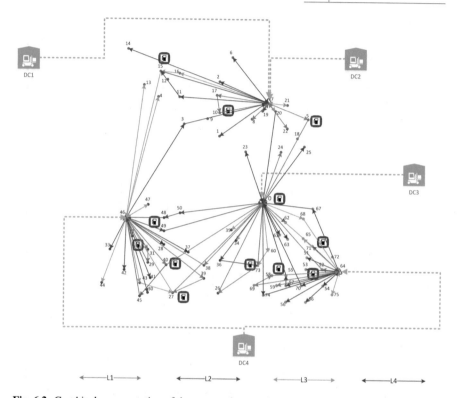

Fig. 6.2 Graphical representation of the proposed network

As given in Table 6.6, the proposed model shares better results compared with current status of the delivery operations in Izmir. In terms of the economic assessment, the worst solution obtained from the proposed model gives around 10% better solution (from 4.6 to 4.2 mil. Euro) than the current status and 138% worse status than the worst scenario in terms of the carbon emissions (from 28.9 to 68.9 kg CO_2 e). The main reason behind these values is the usage of diesel-engine trucks in last-mile deliveries. The carbon emissions rate of the current status (case of truck usage) is calculated by using the equation given by Resat and Turkay [36].

Table 6.5 Individual routes of the drones per different logistics providers in kilometre

Distances [km]				Routes			
L1	L2	L3	L4	L1	L2	L3	L4
Green Zone 1							
4.4	6.1	1.4		7-5-7	7-6-7	7-21-7	
4.6	4.4	3.0		7-22-7	7-2-7	7-20-7	
4.3	3.0	14.0		7-8-7	7-19-7	7-16-15-46	
6.1	5.1			7-17-10-7	7-1-7		
	6.5				7-14-7		
	8.7				7-11-15-7		
Green Zone 2							
8.9	**8.9**	**13.9**	**9.4**	46-39-27-46	46-3-7	46-4-46	46-40-45-46
4.2	2.0	**19.8**	5.1	46-40-46	46-33-46	46-13-46	46-49-0-50-48-46
3.2	4.2	1.4	1.7	46-31-29-46	46-42-46	46-47-46	46-37-46
7.3		5.5	1.0	46-30-46		46-44-46	46-28-46
		16.5				46-43-27-46	
		0.4				46-32-46	
Green Zone 3							
10.3	**10.6**	1.3	**10.3**	0-24-0	0-25-0	0-35-0	0-23-0
13.6	4.2	1.8	5.1	0-38-46-39-27-0	0-36-73-0	0-34-0	0-50-48-46-49-0
7.9	1.2	3.7	**9.3**	0-73-26-0	0-62-0	0-60-73-0	0-74-64
		5.7	1.3			0-68-64-65-0	0-61-0
			0.9				0-63-0
Green Zone 4							
3.2	3.1		4.1	64-67-0	64-71-64		64-70-0
2.6	2.2		2.0	64-52-0	64-51-64		64-54-64
3.0	1.0			64-69-58-64	64-53-64		
2.2	2.0			64-59-64	64-57-55-0		
3.5	6.1			64-75-64	64-56-64		
	5.6				64-66-64		

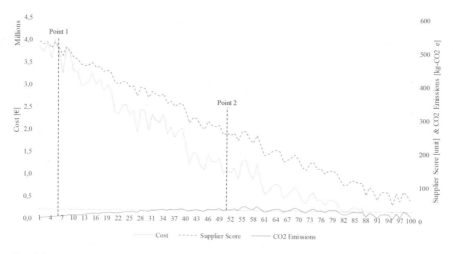

Fig. 6.3 Pareto solution sets for different objective functions

Table 6.6 Solution sets of the proposed model in terms of three objective functions

	f_1 [mil. Euro]	f_2 [unit]	f_3 [kg-CO$_2$ e]
Current status	4.6	529.8	68.9
Single objective	4.2	552.4	2.2
Multi-objective			
Point 1	3.7	509.9	5.7
Point 2	1.1	258.2	28.9

6.6   Conclusion

As sustainability concept is getting much more importance in global competition conditions, companies propose innovative and challenging solutions for daily life operations. Last-mile cargo deliveries occupy the highest importance in logistics activities. Therefore, any improvement or solution methodology may lead to significant improvements in financial, environmental and social aspects. Drone delivery systems or drone routing problems are most popular and highly appreciated concepts in current conditions. This study focuses on novel solution methodology for hybrid drone vehicle routing problem in an aspect of sustainability. The proposed system firstly integrates opinions of many stakeholders (final customers) in the assessment of the logistics providers operating in last-mile delivery systems. After collecting enough data for the proposed model, hybrid solution methodology including TOPSIS and MILP optimization tools takes and processes data sets. At the end of system, different Pareto thresholds are obtained for decision makers, and sensitivity analysis of the solutions is shared. In addition, one of the obtained solutions is compared with the current status of the delivery system in the city of Izmir, and improvements are shared in terms of cost and GHG emissions.

The decision makers can use this model in a ranking of the logistics providers in last-mile deliveries by integrating feedback of the customers under different performance indicators. In addition, obtained Pareto results will enhance the decision-making processes, and stakeholders may get some initiatives or priorities to support environmental conditions in city centres.

One of the expected studies in the near future is inclusion of all characteristics of the drones to create more realistic case and enlargement of the constraints to cover more operational restrictions (such as topographical conditions, climatic conditions (wind, etc.), different type of drone usage, etc.).

References

1. Zampou, E.: U-TURN Project (2018). http://www.citylab-project.eu/presentations/180423_Brussels/4Zampou.pdf. Access time: 22 July 2019
2. CIVITAS: Smart choices for cities making urban freight logistics more sustainable (2019). https://civitas.eu/sites/default/files/civ_pol-an5_urban_web.pdf. Access time: 22 July 2019
3. Lin, C., Choy, K.L., Ho, G.T., Chung, S.H., Lam, H.Y.: Survey of green vehicle routing problem: past and future trends. Expert Syst. Appl. **414**, 1118–1138 (2014)
4. Bektas, T., Laporte, G.: The pollution-routing problem. Transp. Res. Part B Methodol. **45**(8), 1232–1250 (2011)
5. Conrad, R.G., Figliozzi, M.A.: The recharging vehicle routing problem. In: IIE Annual Conference. Proceedings?, p. 1. Institute of Industrial and Systems Engineers (IISE), Atlanta (2011)
6. Erdogan, S., Miller-Hooks, E.: A green vehicle routing problem. Transp. Res. Part E **48**(1), 100–114 (2012)
7. Schneider, M., Stenger, A., Goeke, D.: The electric vehicle-routing problem with time windows and recharging stations. Transp. Sci. **484**, 500–520 (2014)
8. Murray, C.C., Chu, A.G.: The flying sidekick traveling salesman problem: optimization of drone-assisted parcel delivery. Transp. Res. Part C Emerg. Technol. **54**, 86–109 (2015)
9. Ponza, A.: Optimization of drone-assisted parcel delivery. Universita Degli Studi Di Padova Facolta Di Ingegneria Corso Di Laurea In Ingegneria Gestionale (2016)
10. Ferrandez, S.M., Harbison, T., Weber, T., Sturges, R., Rich, R.: Optimization of a truck-drone in tandem delivery network using k-means and genetic algorithm. J. Ind. Eng. Manage. **92**, 374–388 (2016)
11. Wang, X., Poikonen, S., Golden, B.: The vehicle routing problem with drones: several worst-case results. Optim. Lett. **114**, 679–697 (2017)
12. Carlsson, J.G., Song, S.: Coordinated logistics with a truck and a drone. Manage. Sci. **649**, 4052–4069 (2017)
13. Dorling, K., Heinrichs, J., Messier, G.G., Magierowski, S.: Vehicle routing problems for drone delivery. IEEE Trans. Syst. Man Cybern. Syst. **471**, 70–85 (2016)
14. Agatz, N., Bouman, P., Schmidt, M.: Optimization approaches for the traveling salesman problem with drone. Transp. Sci. **524**, 965–981 (2018)
15. Wang, Z., Sheu, J.B.: Vehicle routing problem with drones. Transp. Res. Part B Methodol. **122**, 350–364 (2019)
16. Karak, A., Abdelghany, K.: The hybrid vehicle-drone routing problem for pick-up and delivery services. Transp. Res. Part C Emerg. Technol. **102**, 427–449 (2019)
17. Kitjacharoenchai, P., Ventresca, M., Moshref-Javadi, M., Lee, S., Tanchoco, J.M., Brunese, P.A.: Multiple traveling salesman problem with drones: Mathematical model and heuristic approach. Comput. Ind. Eng. **129**, 14–30 (2019)

18. Mavrotas, G.: Effective implementation of the ϵ-constraint method in multi-objective mathematical programming problems. Appl. Math. Comput. **2132**, 455–465 (2009)
19. Mavrotas, G., Florios, K.: An improved version of the augmented ϵ-constraint method (AUGMECON2) for finding the exact Pareto set in multi-objective integer programming problems. Appl. Math. Comput. **21918**, 9652–9669 (2013)
20. Fattahi, A., Turkay, M.: A one-direction search method to find the exact non-dominated frontier of bi-objective mixed-binary linear programming problems. Eur. J. Oper. Res. **2662**, 415–425 (2018)
21. Velasquez, M., Hester, P.T.: An analysis of multi-criteria decision making methods. Int. J. Oper. Res. **102**, 56–66 (2013)
22. Noorizadeh, A.: Green supplier selection via Multiple Criteria Data Envelopment Analysis (2014)
23. Kannan, D., Khodaverdi, R., Olfat, L., Jafarian, A., Diabat, A.: Integrated fuzzy multi criteria decision making method and multi-objective programming approach for supplier selection and order allocation in a green supply chain. J. Clean. Prod. **47**, 355–367 (2013)
24. Jolai, F., Yazdian, S.A., Shahanaghi, K., Khojasteh, M.A.: Integrating fuzzy TOPSIS and multi-period goal programming for purchasing multiple products from multiple suppliers. J. Purch. Supply Manage. **171**, 42–53 (2011)
25. Ku, C.Y., Chang, C.T., Ho, H.P.: Global supplier selection using fuzzy analytic hierarchy process and fuzzy goal programming. Qual. Quant. **444**, 623–640 (2010)
26. Boran, F.E., Genc, S., Kurt, M., Akay, D.: A multi-criteria intuitionistic fuzzy group decision making for supplier selection with TOPSIS method. Expert Syst. Appl. **368**, 11363–11368 (2009)
27. Liu, P., Zhang, X.: Research on the supplier selection of a supply chain based on entropy weight and improved ELECTRE-III method. Int. J. Prod. Res. **493**, 637–646 (2011)
28. Koksalan, M., Ozpeynirci, S.B.: An interactive sorting method for additive utility functions. Comput. Oper. Res. **369**, 2565–2572 (2009)
29. Koksalan, M., Ulu, C.: An interactive approach for placing alternatives in preference classes. Eur. J. Oper. Res. **1442**, 429–439 (2003)
30. Venkatesan, P., Kumanan, S.: Supply chain risk prioritization using a hybrid AHP and PROMETHEE approach. Int. J. Serv. Oper. Manage. **131**, 19–41 (2012)
31. Rai, D., Kumar, P.: Instance based multi-criteria decision model for cloud service selection using TOPSIS and VIKOR. Int. J. Comp. Eng. Technol. **7**, 78–87 (2016)
32. Resat, H.G.: Design and Analysis of Novel Hybrid Multi-Objective Optimization Approach for Data-Driven Sustainable Delivery Systems. IEEE Access. **8**, 90280-90293 (2020)
33. Stolaroff, J.K., Samaras, C., O'Neill, E.R., Lubers, A., Mitchell, A.S., Ceperley, D.: Energy use and life cycle greenhouse gas emissions of drones for commercial package delivery. Nat. Commun. **91**, 409 (2018)
34. Coelho, B.N., Coelho, V.N., Coelho, I.M., Ochi, L.S., Zuidema, D., Lima, M.S., da Costa, A.R.: A multi-objective green UAV routing problem. Comput. Oper. Res. **88**, 306–315 (2017)
35. CPLEX: IBM ILOG CPLEX optimizer. Int. Bus. Mach. Corp. **46**(53), 157 (2009)
36. Resat, H.G., Turkay, M.: A discrete-continuous optimization approach for the design and operation of synchromodal transportation networks. Comput. Ind. Eng. **130**, 512–525 (2019)

Chapter 7
When Green Technology Meets Optimization Modeling: The Case of Routing Drones in Logistics, Agriculture, and Healthcare

Malick Ndiaye, Said Salhi, and Batool Madani

Abstract The introduction of new green technologies such as electric vehicles, autonomous vehicles, the internet of things, and drones has created new opportunities for improving traditional operations in many industries. In the logistics sector, companies are leveraging on drone technologies to introduce new last-mile delivery solutions. In the agriculture industry, drones are used to survey farming fields for data collection to help improve crop yield or determine areas that require spraying pesticides. These new green technologies can tackle the environmental challenges such as reducing the emission of greenhouse gasses. However, such technologies bring operational challenges, allowing opportunities for researchers to resolve the associated challenges. In this work, we highlight the importance of the drone technology as a green solution in the areas of logistics, agriculture, and healthcare. A review of the recent applications of drone technology is presented to derive the resulted operational challenges. At last, vehicle routing problem variants are presented as potential solutions for addressing the operational challenges of drone technology.

7.1 Introduction

Drone technology is a green solution and an enabler for improving the traditional operations in different fields. From a technology point of view, drones' flying capacities are quite limited compared to the extent of the potential applications. A

M. Ndiaye (✉) · B. Madani
Industrial Engineering Department, American University of Sharjah, Sharjah,
United Arab Emirates
e-mail: mndiaye@aus.edu; g00050500@aus.edu

S. Salhi
Centre for Logistics & Heuristic Optimization (CLHO), Kent Business School,
University of Kent, Canterbury, United Kingdom
e-mail: s.salhi@kent.ac.uk

© Springer Nature Switzerland AG 2020
H. Derbel et al. (eds.), *Modeling and Optimization in Green Logistics*,
https://doi.org/10.1007/978-3-030-45308-4_7

127

drone's flying time also depends on the carried load, the power, the speed, and flying conditions. This usually results in limiting the average flight time to half an hour for commercial drone though this could be made relatively longer in the future [1, 2]. So, to extend the operational time, drones are supported by moving vehicles that may carry extra batteries, parcels, pesticides, or serve as a relay for data collection. As a result, new vehicle routing problems have emerged. When combined with a truck for delivery, the truck acts as a depot for the drone, unlike in traditional routing problems where the depots are fixed. Here, the depot is considered to be moving, which leads to a new challenging delivery and routing problem in which both the drone routes and the location of the truck stops need to be optimized [3]. The problem fits into the class known as location-routing where both problems, namely location and routing, are interrelated [4].

An unmanned aerial vehicle (UAV) is defined as an aircraft with the capacity to fly autonomously due to the support of on-board computers and sensors [5]. It is commonly known as a drone. It has the advantage of being flexible, fast, and easy to move from one point to another. There are several factors that characterize the drone technology, which includes the following as highlighted in [5]:

- The drone's type: determines the shape and the appearance of the drone.
- Drone's level of autonomy: varies from being fully autonomous to being controlled entirely by a remote pilot.
- Drone's size: varies from tiny drones to commercial aircrafts.
- Drone's weight: varies from grams to hundreds of kilograms.

The hardware, software, and network communication performances and capabilities of drones are continuously improving. Drones' lithium batteries are taking a significant share in these enhancements to extend the distance traveled by the drone on a single charge. In addition, innovations are made on the drone's software used for tracking and navigation, including the operating systems that are used to optimize the drone's routes to avoid unfavorable weather conditions and risk factors [6]. Given that drones operate in a 3D environment, sensors to detect obstacles that can either fixed or moving and be able to adjust the speed based on the wind direction and strength need to be sophisticated. This extra complexity makes the routing of drones even more challenging.

Drones have existed for almost a century. However, with the intense focus today on its advanced technology and people's awareness of the environmental impact, drones are today small, inexpensive, and readily available. They are potentially considered to be a powerful and cost-effective tool. They are useful for tasks that are dull, hazardous, and dirty [7]. For instance, its use is crucial and life-threatening when it comes to delivering goods and medicine to remote and affected areas by natural as well as man-made disasters. Thus, drone technology has gained significant interest in using it in different sectors and fields. The new technology has attracted media attention. As a result, various industries are considering the adoption of drones. As an example, Amazon has sold more than 10,000 drones within a month in 2014 [8]. Besides the low acquisition cost and the ease of availability, the use of drones also has a positive impact on economical and environmental sustainability. Ideally, drones yield more economical energy consumption and decrease greenhouse gas emissions, thus reducing the carbon footprint while enhancing environmental sustainability [7].

The chapter provides an overview of the characteristics of drones' technology and their impacts on designing more sustainable operations in logistics, agriculture, and healthcare. The review of the relevant industrial applications provides insight on new routing challenges that lead to new vehicle routing problems such as the Drones Vehicles Routing Problems (DVRP) and hybrid systems when combined with other delivery logistics modes. We also offer some insights into the possible models as well as the research potential for future work, mainly focusing, on this occasion, on the last mile delivery in the area of green logistics. Possible adaptations to other related combinatorial optimization problems will also be briefly discussed.

7.2 Drone as a Green Technology

Transportation, including last-mile delivery, contributes mostly to the total greenhouse gas emissions in the logistics industry [9]. With e-commerce being expected to grow daily, the pressure on the logistics providers to tackle the challenges of conveying goods from transportation hubs to the final destination in the supply chain management [10] is known as the last-mile problem. Logisticians consider it the most polluting, most expensive, and the least efficient part of the supply chain.

From the sustainability point of view, the vast number of daily deliveries, including the repeated deliveries, has negatively contributed to the increase in the emission of air pollutants (CO_2). This is due to the unavailability of customers at their place of residence resulting in the increment of the car mileage of the courier. It is found that approximately 20%–30% of deliveries are consumed by additional mileages to revisit the same customers [11]. Thus, transportation activities are associated with the increase of the levels of environmental externalities. According to Rodrigue et al. [12], 15% of global carbon dioxide emissions attribute to the transport sector. To tackle this pollution challenge, governments are encouraging the companies to replace their diesel/petrol delivery vehicles with a battery-powered car. This shift will contribute to improving the last-mile delivery process, as they can cut down on the number of small road deliveries and reduce the number of traditional vehicles on the road. It is worth stressing that the energy needed to burn the crude oil to diesel adds an extra 20% of greenhouse gasses. Besides, it was discovered that burning a gallon of diesel emits approximately 10 kg of carbon dioxide [13].

However, drones have a limited load capacity that restricts their use to the delivery of small packages only or dropping off/picking up emergency supplies, vaccines, and medicines. In other words, drones are more convenient than trucks for small deliveries and short traveling distances. With the continuously evolving technology, the drones will eventually have, in the near future, larger load capacities and longer traveling distances, which will result in a significant reduction in the amount of greenhouse gas emissions.

The benefits of using drones as a greener transportation mode also extend to other areas such as agriculture, including crop monitoring fields. They are utilized as precision agricultural instruments, replacing the gas-guzzling machines to reduce the amount of fertilizer use by about 20% while protecting the environment from pollution emitted in the process [14]. Drones are also found to be an efficient way to improve the water requirement and the number of harmful chemicals. There are 3–5% of crops often lost due to sprayers and heavy equipment entering the farming fields, which is entirely avoided when drones are adopted instead. Besides, farmers use drones in monitoring livestock, crops, water levels, vegetation growth, and soil health as well as providing high-resolution images to inform in detail about crop health. Drones with high specifications can also create three-dimensional images of the landscape for future expansions [15].

7.3 Drone: Industrial Applications

Drone's applications involve delivering small loads, surveillance using on-board cameras, crop monitoring, land spraying, scientific research, emergency services, and disaster response. Given the above-mentioned benefits, in this section, we look further at the applications of drones in logistics, healthcare, and agriculture. The implicit routing considerations are discussed in the next section.

7.3.1 Logistics

Drones have attained great attention in the logistics systems to tackle the challenges of last-mile delivery problems resulting from the online shopping platforms. The delivery service offered by e-retailers is one of the fundamental factors influencing the customers' decision to shop from them, so the ability to deliver customers' orders on time reflects the success of businesses. Thus, logistics providers are working to improve their traditional delivery operations using autonomous vehicles, including drones. Companies such as DHL, Amazon, Google, and UPS have already announced the use of drones in their delivery systems. For example, UPS subsidiary UPS Flight Forward Inc. has received the U.S. government's first full Part 135 Standard certification to operate a drone airline in campus settings such as hospitals and universities but not in residential areas yet [16].

DHL is giving a significant attention to the applications of drones stating that electrical drones appear to be the most promising type of drone for e-commerce systems [17]. They have tested the implementation of drones in rural areas (mountain areas), as a supporting system to their parcel lockers. Parcelcopter Skyport is designed to operate by inserting first the package into the parcel locker. Then the drone swoops into the action through a helipad on top of the locker, grabs the package, and deliver it. The drone is capable of flying 70 km per hour, making deliveries within 8 min, and carrying up to 5.5 pounds for 8.3 km [18].

Amazon is one of the leading companies in the implementation of drones for delivery purposes. They have attained patents on different designs such as fulfillment center towers with drone delivery, in which the towers are served as drone charging and packages pickup hubs. Multiple drones are launched from the tower for the delivery of multiple customers [19]. Also, Amazon airborne fulfillment center for drones suggests that drones are released from the airborne to customers. Besides drones, smaller airships (shuttles) are also used to replenish airborne and drones with the necessary inventory and fuel [20].

A challenging new research avenue would be to combine drones with traditional delivery methods such as trucks to form a hybrid truck–drone system. Drones are known to have limited traveling distances and limited load capacity. In contrast, trucks have usually a relatively long-range of travel capability and can transport large and heavy packages. However, trucks are heavy, slow, have high transportation costs, and have a significant amount of greenhouse gas emissions besides not being able to access poorly connected road networks in some harsh areas. Consequently, these benefits of trucks and drones can then be taken advantage of rendering their combination worthwhile exploring [21].

Hybrid truck–drone systems are now attracting not only academics but also companies. According to UPS Company, which is testing this system, drone delivery can help in lowering the cost, specifically in rural locations where cars must drive miles between single deliveries. It was reported that this system could save up to $50 million per year by cutting a mile off in every driver's route each day [22]. However, aviation administration has generally prohibited commercial drones from flying beyond the sight of their pilots [23]. Mercedes Benz is also investing $562 million into delivery van–drone. The van, with a range of up to 168 miles, has a fully automated cargo space and consists of a mechanical shelving system that can load packages and autonomously identify the package's destination. The driver gets a notification when reaching a drop-off area, and the shelving system will push the package to the drone that will deliver it to the customers [24].

Furthermore, a train-mounted hub for drone delivery is a patent received by Amazon, where the drone can pick up packages and return to the train for charging or further pickups while the train is in motion [25]. Table 7.1 summarizes the technical characteristics of current applications of drones in the area of logistics.

7.3.2 Agriculture

Efficiency and flexibility are critical for crop protection. On-time protection minimizes losses from insects, diseases, and others to increase the yield quality. Drones are essential in agricultural operations due to their low cost and high efficiency. Using drones in surveying services and providing maps helps farmers in understanding their fields and in making plans for the next growing seasons [28].

Table 7.1 Applications of drones in Logistics

Technology	Characteristics
Prime air delivery drone (Amazon)	• Can deliver orders under 2.3 kg to customers within a 16 km radius of Amazon's fulfillment centers [26]. • Navigates through the on-board global positioning system (GPS) [27]
Parcelcopter Skyport (DHL)	• Can fly 70 km/h for 8.3 km, and it can do deliveries within 8 min and tested in mountain areas. • Packages are inserted in the parcel locker, drone swoops into action (through a helipad on top of the locker), grab the package, and deliver [18].
Fulfillment center towers with drone delivery (Amazon)	• Towers are served as drone charging hubs and allow drones to pick up packages [19].
Train-mounted mobile hubs for drone delivery (Amazon)	• Drones can pick up packages and return for charging or further pickup while the train is in motion [25].
Amazon airborne fulfillment center	• Unmanned aerial vehicles (UAVs) are released from the airborne to deliver to designated customers. Smaller airships (shuttles) are used to replenish airborne and UAVs with inventory and fuel [20].

Drones are small, fast, and flexible, which provide them the ability to perform different functions in the agriculture field. For instance, drones can be employed to deliver essential materials to crops. Also, for crop dusting, drones can be used instead of planes. The following are some of the potential employment of drones in agriculture:

• Planting: the drones distribute pods seeds and plant nutrients into the soil to provide the plant with all the necessary nutrients to sustain life. Start-ups have decreased planting costs by 85% using drones for planting.
• Crop spraying: scanning the ground and spraying the correct amount of liquid can be achieved by drones. According to experts in this field, spraying using aerial vehicles can be up to five times faster than the traditional method. This process will undoubtedly increase the efficiency of reducing the number of chemicals penetrating the soil. In Japan, farmers use autonomous helicopters, and 40% of their rice fields are now being sprayed using drones. Similarly, AeroDrone is

a Ukraine-based start-up project for a fixed-wing spraying UAV. Comparing to traditional crop protection methods such as aircraft and tractors, AeroDrone has a low fuel consumption, high productivity, low noise pollution, and no chemical contamination risks for operators [29].

- Crop monitoring: to monitor the field, images of the crop field should be captured on a regular basis. Using satellite imagery is useful; however, it has different drawbacks. For instance, the Normalized Difference Vegetation Index (NDVI) is used to identify the farmlands of poor soil fertility. Satellite images are used to calculate the index. However, they are costly, time-consuming, and can easily lead to poor management decisions. Hence, it is reasonable to use drones for crop monitoring as they are cheaper and equipped with sensors such as Near Infrared. Therefore, drones can replace the traditional way and provide precise images of the field to define any inefficiencies with no time delay [29].
- Irrigation and field analysis: instead of irrigating the entire field, as commonly used, drones are equipped with a technology that identifies the dry parts using specific sensors. Besides irrigation, drones can produce three-dimensional maps that are useful in planning seed patterns as well as providing data for irrigating and nitrogen-level management [30]. At Timiryazev State Agrarian University in Moscow, a research team used a farming drone for capturing high-resolution imagery for the wheat fertilization project. The imagery was used to create a custom application map to optimize the nitrogen application, leading to a 20% reduction in nitrogen [31].

The agriculture field pays major attention to drone technology, as it provides a high-technology transformation in planning, data gathering, and solutions. This can be concluded from the continuous improvements in the technology of agricultural drones. The implementation of drone in agriculture is around the globe with different characteristics and specifications. Amazon is interested in developing drones to support the agriculture industry. Hybrid Airship Drone Farm Robot system is a potential design by Amazon that is used for crop dusting, planting, and fertilizing. The system is supported by a gas balloon and propellers for lifting. Comparing to other drones, these are large making the path planning relatively more challenging. The farm robot is restricted to start from the base station, completes its service, and returns to the same station. The length of the path traveled by the robot is estimated based on the application intensity [14]. Table 7.2 summarizes the use of drone technology in agriculture including monitoring, spraying, and field analysis [32].

7.3.3 Healthcare

Drone technology plays a substantial role in delivering and picking up medications, vaccines, and blood samples, especially in rural areas with poor infrastructure and challenging geographical conditions.

Table 7.2 Applications of drones in agriculture

Technology	Characteristics
AgDrone (Honeycomb)	• Considered as a sophisticated drone. • Covers 600–800 acres of every hour and flies at 400 feet. • Flies at a maximum speed of 82 km/h. • Supported by an automatic dual-camera electrical signal.
Matrice 100 (DJI)	• Considered as the best quadcopter-based drone for agriculture. • Contains dual battery support. • Supported by GPS and flight controller. • Can operate in all environmental conditions (operating temperature is -10–$40°$ C).
T600 Inspire 1 (DJI)	• Is a quadcopter-based drone for agriculture. • Contains individual flight and camera control and 4K video recording. • Has easy navigational capabilities. • The maximum flying time is between 18 and 40 min.
Agras MG-1 (DJI)	• Used to help farmers in spraying large areas of farms with pesticides, fertilizers, etc. • Can array 10 kg of liquid payload. • Can cover 1–1.50 acres within 10 min. • Operates under temperatures between 0 and $40°$ C.
EBEE SQ (SenseFly)	• Is designed for crop monitoring and identifying problems during flight. • Is provided with a multispectral sensor used to capture data. • Has automatic three-dimensional flight planning. • The maximum flying time is 55 min.
Pam 20 (AuroDrone)	• Has a suspended portable control unit that includes measuring devices, sensors, and other electronics. • Can carry up to 25 kg of payload weight. • Can cover 250 km at a maximum speed of 80 km/h. • The maximum flying time is 3 h.

San Francisco-based drone start-up, Zipline, transports blood for transfusion across Rwanda [33], completing about 150 blood deliveries per day to transfusing facilities on the western side of the country. To do so, the company has overcome multiple challenges, such as designing a drone that is capable of traveling for long distances. Likewise, Matternet has developed a drone that can carry 2.2–4.4 pounds and move items about 10 kilometers, with a speed of up to 40 kilometers per hour and a flying time of 18 min. The developed drone has delivered medications in Haiti following the 2010 earthquake and in the Dominican Republic, Switzerland, and Guinea. The drone operates as follows: a smartphone is used to choose from the list of designated destinations, and then the drone forms a route based on the weather, population density, and airspace to avoid schools, public spaces, among others [34].

Moreover, prototype ambulance drones have demonstrated a distribution of defibrillators in the Netherlands. The drones are estimated to reach the patients within a 4.6 square mile radius in a minute versus an average of 10 min for traditional emergency services. They travel up to 60 miles per hour, allowing a fast response, which increases the possibility of survival to 80% versus 8% for traditional emergency services. This is an exciting and promising result with a massive health and societal impact.

Furthermore, a drone called Flirtey is the first drone approved by the Federal Aviation Authority for dropping off medical supplies to a health center in Virginia rural side. The drone has succeeded in delivering prescription items from the Wise County Regional airport in approximately 3 min. In contrast, a regular truck delivery would accomplish the same task within 90 min, approximately 95% slower. Also, Flirty drones performed deliveries of aid kits and emergency medications in Nevada, Australia, and New Zealand [33].

Google is also interested in using drone technology for transporting medications. It received a patent that outlines a system where drones bring medical aid to people who need supplies urgently, where an ambulance cannot make it at the needed time. However, to guarantee patient safety and health regulations, when the drone is on its way, it also sends a support request to ensure the ambulance in on its way. Drones have the flexibility to adapt to any situation. For instance, it can land on the water like a floatplane if needed. The system is designed to work in the following way: the person who needs help can call the drone using a portable box that allows entering the type of medical problem. Then, the drone will address the medical problem selected in the portable box and deliver the needed supplies. When delivering, the drone will provide instructions on how to use the supplies via video [33].

Table 7.3 presents some of the recent applications of drones in the sector of healthcare, showing the company developing the drone and its corresponding operating characteristics.

Table 7.3 Applications of drones in healthcare

Technology	Characteristics
Seattle's VillageReach (Seattle and MetterNet)	• Used to transport blood samples across hospitals. • Can carry up to 2 kg and cover a range of 10 km with a speed of 25 mph [33].
TU Delft (DHL)	• Is supported by a cardiac defibrillator and a two-way communication radio with video. • Can travel up to 12 km at a speed of 60 mph [6].
Zipline	• Used to transport blood for transfusion across the country. • Can carry up to 3 lb and cover a range of 45 miles with a speed of 90 mph [6].
EHang	• Is used to automating the transportation of organs across the country in emergency situations. • Is capable of carrying a single person with a maximum weight of 210 kg at a speed of 130 kph [33].

7.4   Routing Considerations

The drones' systems are associated with distinguished path planning and routing considerations. This conclusion is derived from the different applications mentioned above, in which drone systems have various factors that would impact the design of the routing problems. These factors include the number of deliveries per route, number of routes, number of drones, and depot position (fixed or moving). For instance, the train-mounted hub for drone delivery system deals with a single drone that can carry single delivery per route. So, for the drone to deliver more than one package, the construction of multiple routes is needed. In addition, the train is acting as a moving depot; that is, the drone is released and collected by the train when it is moving. Thus, routing consideration should account for the drone and the train movements. In addition, drones used in the healthcare sector such as Flirtey drones are used for picking up and deliveries blood samples. In such a case, drones can make either a single route or multiple routes depending on the number of blood samples that need to be delivered limited by the load capacity of the drone. Table 7.4 summarizes the routing consideration of drone systems, emphasizing their applications as well as their characteristics in terms of capability and limitations. In this section, we will also discuss some of these works with supporting examples from the literature. This is followed by a brief on the formulations that arise for the vehicle routing problems associated with drones and hybrid truck–drone systems.

7.4.1   A Brief Review of Drones on the Last Mile Delivery

With drone technologies being implemented in different industries and fields, operational challenges occur. One challenge is the routing consideration of the drone. The routing decisions are constrained by the application of the drone and its technical specifications such as a battery, payload capacity, maximum flying time, cost, speed, etc. In addition, routing decisions can be dynamic and real time-dependent. In distribution networks, for instance, the routing decision would be dynamic if the system redistributes resources in real time, depending on the prioritization of the shipments. Similarly, in transporting blood, the transfer of blood can be sequenced depending on its urgency and time criticality. In addition, picking up packages by drones is dynamic when the customer can select the closest drone to his/her collection location [3]. The continuous evolvement of drone technology has motivated researchers to focus on its routing decisions in different sectors such as logistics, agriculture, and healthcare. In logistics, for instance, the focus is on the last-mile delivery problem, which we aim to cover in the rest of the section.

Last-mile delivery is the least efficient part of the e-commerce supply chain. It accounts for 13%–75% of the total supply chain cost [35]. As stated earlier, combined truck–drone system is considered a potential solution for such problem. Murray and Chu [36] suggested the flying sidekick system that integrates the use of

Table 7.4 Routing considerations in drone technology

System	Application	Number of deliveries/route		Number of routes		Number of drones		Depot position	
		Single	Multiple	Single route	Multiple routes	Single Drone	Multiple Drones	Fixed depot	Moving depot
Drone Systems	Prime air delivery drone	✓			✓	✓		✓	
	EBEE SQ drone	✓	✓	✓	✓	✓		✓	
	Flirtey drone	✓	✓	✓	✓	✓		✓	✓
Hybrid systems	Fulfillment center towers with drone delivery	✓					✓	✓	
	Train-Mounted mobile Hubs for drone delivery	✓			✓	✓			✓
	Amazon airborne fulfillment center	✓			✓		✓		✓
	Hybrid airship drone farm robot system		✓	✓		✓		✓	

one drone and one truck. The distribution of packages works as follows: customers are served once either by the truck or by the drone depending on the parcel size. If the drone cannot transport the package, then the truck will deliver it. From a routing point of view, the combined system departs and returns to a single point only once, either the depot or customers' nodes. Also, to accommodate the drone's battery consumption, the truck might transport the drone and return to the depot or for further customers' deliveries.

Similarly, Joeng et al. [37] developed a truck–drone hybrid delivery system for delivery purposes, which takes into consideration the impact of parcel weight on drone energy consumption, drones' limited batteries, and the restricted flying areas of drones. As in Murray and Chu's paper, the drone is only launched and collected by the truck at either the customers' nodes or depots. On the other hand, Karak and Abdelghany [38] developed a similar system but with the ability to pick up and deliver. The system focuses on minimizing the total routing costs with the inclusion of a drone's flying range and load capacity. Furthermore, drones can be used to resupply the delivery vehicles with the required packages. This innovative system was suggested by Dayarian et al. [39], in which they investigate the resupplying configurations such that the resupply takes place when the truck is not in motion. The drone's handover of the package occurs at the roof of the vehicle.

For agriculture, recent studies have focused on optimizing the drone's task in spraying nutrients and other needed materials. Luo et al. [40] studied the effect of using drones in pesticide spraying through optimizing the entire process from determining the drones' optimal paths to allocating the tasks for the drones. Two effects were taken into account, the type of pesticide and the temperature during spraying. The designed drones can automatically avoid obstacles by using the control strategy of self-circumvention. Also, they have constant speeds to avoid any distractions in the spraying process.

Lastly, in the healthcare sector, Kim et al. [41] designed delivery and pickup system for rural areas, where patients have to visit clinics for health testing and medicine fill-up. The drone is specified for a load capacity of a maximum of three deliveries and pickups per route. The objective is to minimize the operational cost of the drone resulting from delivering medicines and picking up exam kits, as well as optimizing the number of visited locations per route. Additionally, Scott and Scott [6] provide a model for m-trucks and m-drones systems to minimize the total weighted delivery times with taken into account budget constraints. Drones are assumed to carry one item at a time, can cover 100 miles per delivery trip, and have constrained battery. Also, drones are assumed to fly in a straight line, in which distances are calculated using the Euclidean metric. On the other hand, trucks can make multiple deliveries at a time and follow the road network, in which distances are calculated empirically or using GIS.

7.4.2 Potential Formulations for Drone Routing

We provide two vehicle routing problems associated with the use of drones. One is based on the drones only, while the other combines drones and trucks in an integrated system.

1. **Simple drone routing**

    The routing considerations for a drone system is originated from the traveling salesman problem. The traveling salesman problem (TSP) is one of the well-known problems in combinatorial optimization and refers to a salesman who is to visit a set of cities and return to the initial city (depot) with the objective to minimize the total distance traveled [42]; the order is called a tour or circuit through the cities [43]. Many researchers have addressed it under various schemes. Their model solutions range from exact methods where optimality can be guaranteed to heuristics and metaheuristics, which are faster but do not guarantee optimality (for an overview of this aspect, see Salhi [44]).

    The following are the parameters and decision variables commonly used to describe the TSP mathematical model alongside the mathematical formulation developed by Desrochers and Laporte [45].

    Parameters
    n Number of nodes
    c_{ij} Cost matrix (distance) between node i and node j $(i, j = 1,\ldots,n)$

    Decision Variables

    $$x_{ij} \quad \begin{cases} 1 & \text{if the link exists between node } i \text{ and node } j \\ 0 & \text{otherwise} \end{cases}$$

    u_i Number of nodes visited from the depot to node j

    Mathematical Formulation of the TSP (Drone Routing)

    $$M_{in} = \sum_{i=1}^{n} \sum_{j=1}^{j} c_{ij} x_{ij} \tag{7.1}$$

    $$\sum_{i=1}^{n} x_{ij} = 1, \forall j \tag{7.2}$$

    $$\sum_{j=1}^{n} x_{ij} = 1, \forall i \tag{7.3}$$

    $$u_i - u_j + (n-1)x_{ji} \leq n-2, \qquad \forall i < n, \forall j, i \neq j \tag{7.4}$$

Equation (7.1) gives the objective function that minimizes the total distance to cover all cities. Constraints (7.2) and (7.3) ensure that every city is visited only once. Constraint (7.4) is a sub-tour elimination constraint (SEC).

2. Hybrid Truck–Drone Combined System

Madani and Ndiaye [46] have recently developed a mathematical model that works as follows: a truck moves on a particular path and releases a drone to serve a set of customers. Once all customers are served, the truck collects the drone. The optimization model attempts to obtain the optimum configuration of a set of drone routes and to find the optimal locations for the truck to release and collect the drone in order to minimize the traveling cost required to visit all customers on a single trip. The proposed model is referred to as the traveling salesman problem with a moving depot (TSP-MD), where the following assumptions are considered.

- The truck–drone system departs from a fixed depot, but the drone might be released from either the fixed depot or any point along the path.
- The drone visits all customers and returns to the truck.
- Customers are available during the time of the delivery, and each customer is visited only once.
- Truck and drone's traveling distances are Euclidean.

The following are the problem parameters (some are similar to the ones in the previous case) and variables used to formulate the problem, which is an integer linear problem (ILP).

c_{ij} Distance traveled by the drone between node i and node j
t_{ij} Distance traveled by the truck between node i and node j
n Number of customers' nodes
L, l Set of predetermined locations
F_d Drone unit cost (cost per unit distance)
F_t Truck unit cost (cost per unit distance)

$$x_{ij} \quad \begin{cases} 1 & \text{if the drone travels from } i \text{ and node } j \\ 0 & \text{otherwise} \end{cases}$$

$$y_{ij} \quad \begin{cases} 1 & \text{if the truck travels from } i \text{ and node } j \\ 0 & \text{otherwise} \end{cases}$$

u_i Number of nodes visited from the depot to node j

The mathematical formulation is given below, followed by its description.

$$Min = \sum_{i=0}^{l+n}\sum_{j=0}^{l+n} F_d c_{ij} x_{ij} + \sum_{j=0}^{l} F_t t_{0j} y_{0j} \tag{7.5}$$

$$\sum_{j=l+1}^{l+n} x_{0j} = 1, \tag{7.6}$$

$$\sum_{i=0}^{l+n} x_{ij} = 1, \qquad j = l+1,\dots,l+n \quad i \neq j \tag{7.7}$$

$$\sum_{j=0}^{l+n} x_{ij} = 1, \qquad i = l+1,\dots,l+n \quad i \neq j \tag{7.8}$$

$$\sum_{i=0}^{l} \sum_{j=l+1}^{l+n} x_{ji} = 1, \tag{7.9}$$

$$\sum_{i=0}^{l+n} x_{i0} = 0, \tag{7.10}$$

$$y_{0j} = \sum_{i=l+1}^{l+n} x_{ij}, \qquad j = 0,\dots, L \tag{7.11}$$

$$u_i - u_j + n x_{ij} \leq n - 1, \qquad \forall i > l+1, \forall j, \quad i \neq j \tag{7.12}$$

Equation (7.5) presents the objective function that minimizes the total traveling cost for the truck and the drone. Constraint (7.6) ensures that the drones start its route from the predetermined location only once. Constraints (7.7) and (7.8) are to ensure that customers' nodes are entered and departed from only once. Constraint (7.9) ensures that the drone enters the collection location only once. Constraint (7.10) ensures that the drone does not return to the initial predetermined location. Constraint (7.11) indicates that the variable y_j exists if the drone enters the collection region. Finally, constraint (7.12) is a sub-tour elimination constraint (SEC).

The TSP-MD problem is NP-hard. Indeed, if we restrict the path of the truck to a single location (initial depot), then the routing problem reduces to the finding of a single drone route. The problem is then reduced to a TSP, which is known to be NP-hard [44].

7.4.3   Some Computational Results

We draw several data set from the online TSP library [47] to solve the problem using the General Algebraic Modeling System (GAMS) optimizer software. Table 7.5

Table 7.5 Computational time for the different sets of customers' nodes

Number of Customers' Nodes	Computational Time in Seconds					
	Sample 1	Sample 2	Sample 3	Sample 4	Sample 5	Average
11	0.97	0.59	0.91	0.59	0.98	0.808
21	1.97	0.97	1.83	1.74	1.03	1.508
31	2.22	1.66	2.66	2.45	1.19	2.036
41	4.98	2.83	2.98	26.16	11.17	9.624
51	4.47	4.58	4.75	5.08	101.16	24.008
61	101.13	27.16	17.19	3.63	119.67	53.756
71	261.66	436.25	32.19	6.16	15.91	150.434
81	145.16	384.5	464.52	219.09	384.67	319.588

shows the results for eight different sizes of customers randomly selected from an initial data set of 100 nodes. For each problem size, the problem was solved 5 times for 5 different sets of customers. The average computational time is then recorded, and a plot against the number of customers is given. The experiments show that small instance problems, with $N <= 100$, can be solved optimally within 30 min. However, the computational time increases exponentially with the number of customers as illustrated from Fig. 7.1. These results illustrate the complexity of combining two routing decisions (truck and drone) in the same problem.

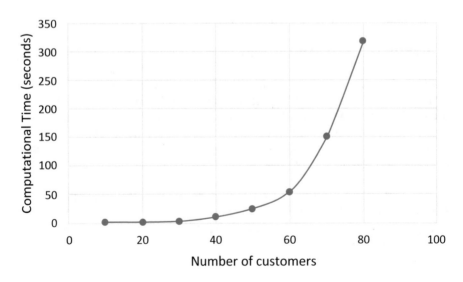

Fig. 7.1 The computational time versus the number of customers

7.5   Conclusion and Potential Research Avenues

We first summarize our review and then present some research avenues that we believe to be worthwhile exploring in the future.

In this chapter, we have highlighted the effect of drones as a green technology and its potential in bringing massive benefits to a sustainable economy. More emphasis is given to three areas, namely, logistics, agriculture, and healthcare, where drones' operations seem to prove to be extremely important, and in some situations, critically vital. The industrial applications of drone technology also open a new challenge in the area of optimization. For instance, when dealing with drone, effective routing, which can be treated separately or in conjunction with other delivery systems such as trains or trucks, is an interesting and practical new problem. Optimization challenges of such hybrid systems have been briefly described, and an attempt to mathematically formulate a simple variant is also provided.

As this green technology will be with us for the future, and as this will help us in resolving the environmental challenge we all face, opportunities for academics in terms of research are immense. For simplicity, we present some aspects that we believe can be worth exploring. One should consider all formulation variants known for traditional VRP such as a time window, restriction on the number of vehicles, fixed charges, variable charges, and varying vehicle capacity. Other challenges, such as battery/power limitations for a drone or adding recharging stops during deliveries, need to be addressed. Developing efficient heuristics solutions to solve the routing of hybrid truck–done systems is also another research avenue that deserves attention. As we combine simultaneously two dependent systems, namely routing and location into one global problem, also known as a location-routing problem in the literature, this becomes an exciting research avenue (see [4]). Solution methods could be based on new developments of mathematical programming, whereas others could focus on developing efficient and powerful metaheuristics for such complex but practically useful logistical and environmentally based problems. Any progress along these lines will not only advance our understanding in dealing with such contemporary problems but also, and more importantly, help us to address an environmentally linked problem with the aim to reduce CO_2 emission and hence improve our health and that of the new generation, namely our children and grandchildren.

Acknowledgments The authors are grateful to the referees' suggestions which improved the content and the presentation of the chapter.

References

1. Lim, J., Jung, H.: Drone delivery scheduling simulations focusing on harging speed, weight and battery capacity: case of remote islands in South Korea. In: *2017 Winter Simulation Conference (WSC)*, pp. 4550–4551 (2017). https://doi.org/10.1109/WSC.2017.8248199
2. Figliozzi, M.: Drones for Commercial Last-Mile Deliveries: A Discussion of Logistical, Environmental, and Economic Trade-Offs (2017)

3. Heutger, M., Kuckelhaus, M., Zeiler, K.: Self- Driving Vehicles in Logistics: DHL. http://www.dhl.com/content/dam/downloads/g0/about_us/logistics_insights/dhl_self_driving_vehicles.pdf
4. Nagy, G., Salhi, S.: Location-routing: issues, models and methods. Eur. J. Oper. Res. **177**(2), 1 (2007)
5. Smith, K.W.: Drone technology: benefits, risks, and legal considerations. Seattle J. Envtl. L. **5**, i (2015)
6. Scott, J., Scott, C.: Drone delivery models for healthcare. In: Proceedings of the 50th Hawaii International Conference on System Sciences (2017)
7. Chiang, W.-C., Li, Y., Shang, J., Urban, T.L.: Impact of drone delivery on sustainability and cost: Realizing the UAV potential through vehicle routing optimization. Appl. Energy **242**, 1164–1175 (2019)
8. Custers, B.H.M.: The Future of Drone Use : Opportunities and Threats from Ethical and Legal Perspectives. Information Technology and Law Series, vol. 27, pp. 2215–1966. Asser Press/Springer, Berlin (2016, in English)
9. Hertwich, E.G., Peters, G.P.: Carbon footprint of nations: a global, trade-linked analysis. Environ. Sci. Technol. **43**(16), 6414–6420 (2009)
10. Yu, J.J.Q., Lam, Y.S.: Autonomous vehicle logistic system: joint routing and charging strategy. IEEE Transactions on Intelligent Transportation Systems (2017)
11. Moroz, M., Polkowski, Z.: The last mile issue and urban logistics: choosing parcel machines in the context of the ecological attitudes of the Y generation consumers purchasing online. Transp. Res. Procedia **16**, 378–393 (2016)
12. Rodrigue, J.-P., Comtois, C., Slack, B.: The geography of transport systems. Routledge, London (2016)
13. Samaras, C., Stolaroff, J.: Is Drone Delivery Good for the Environment? (2018)
14. Salnikov, V., Filin, A., Bahi, S.: Hybrid Airship-Drone Farm Robot. US Patent Application 15/193,033 (2016). https://patentimages.storage.googleapis.com/03/1a/59/434b2b2d63bb8d/US20160307448A1.pdf
15. Ogleby, G., Joshi, C.: Five Ways Drones are Being Used to Help the Environment (2016)
16. Baertlein, L.: Big drone on campus: UPS gets U.S. government okay for drone airline (2019)
17. Koiwanit, J.: Analysis of environmental impacts of drone delivery on an online shopping system. Adv. Clim. Change Res. **9**(3), 201–207 (2018)
18. Franco, M.: DHL uses completely autonomous system to deliver consumer goods by drone (2018). https://newatlas.com/dhl-drone-delivery/43248/. Accessed 02- Jan- 2018
19. Margaritoff, M.: Amazon Patents Fulfillment Center Towers to Increase Drone Delivery Efficiency (2018). http://www.thedrive.com/aerial/11865/amazon-patents-fulfillment-center-towers-to-increase-drone-delivery-efficiency (Accessed 19-Feb-2018)
20. Berg, P., Isaacs, S., Blodgett, K.: Airborne Fulfillment Center Utilizing UAV for Item Delivery (2016). United States Patent Appl. 14/580,046. https://patents.google.com/patent/US9305280B1/en
21. Ha, Q.M., Deville, Y., Pham, Q.D., Hà, M.H.: On the min-cost traveling salesman problem with drone. Transp. Res. C Emerging Technol. **86**, 597–621 (2018)
22. UPS Tests Drone Delivery System (2018). http://www.businessinsider.com/ups-tests-drone-delivery-system-2017-2 (Accessed 06- Jan- 2018)
23. Kastrenakes, J.: UPS has a delivery truck that can launch a drone (2018). https://www.theverge.com/2017/2/21/14691062/ups-drone-delivery-truck-test-completed-video (Accessed 06- Jan-2018)
24. Muojo, D.: Mercedes is reportedly pouring $562 million into delivery van drones—here's a glimpse of what's to come (2018). http://www.businessinsider.com/mercedes-electric-vision-van-drone-delivery-service-photos-2017-3/#meet-the-mercedes-vision-van-an-all-electric-van-with-a-range-up-to\-168-miles-1 (Accessed 06- Jan- 2018)
25. Margaritoff, M.: Amazon Patents Train Mounted Mobile Hubs For its Drone Delivery Fleet (2018). http://www.thedrive.com/aerial/13431/amazon-patents-train-mounted-mobile-hubs-for-its-drone-delivery-fleet (Accessed 19-Feb-2018)

26. Lavars, N.: Amazon reveals Prime Air drone capable of 30 min deliveries (2018). https://newatlas.com/amazon-prime-air-delivery-drone/29982/ (Accessed 03- Jan- 2018)
27. Rose, C.: Amazon's Jeff Bezos looks to the future (2018). https://www.cbsnews.com/news/amazons-jeff-bezos-looks-to-the-future/ (Accessed 03- Jan- 2018)
28. Du, B.: A Precision Spraying Mission Assignment and Path Planning Performed by Multi-Quadcopters. In: *Proceeding of the 2017 2nd International Conference on Electrical, Control and Automation Engineering (ECAE 2017)*. Atlantis Press, New York (2017)
29. Pederi, Y., Cheporniuk,H.: Unmanned aerial vehicles and new technological methods of monitoring and crop protection in precision agriculture. In: Proceedings of the 2015 IEEE International Conference Actual Problems of Unmanned Aerial Vehicles Developments (APUAVD), pp. 298–301. IEEE, New York (2015)
30. Mazur, M.: Six Ways Drones Are Revolutionizing Agriculture (2016)
31. Folk, E.: How Drones are Helping the Environment, and Why That is Important?
32. Puri, V., Nayyar, A., Raja, L.: Agriculture drones: a modern breakthrough in precision agriculture. J. Stat. Manage. Syst. **20**(4), 507–518 (2017)
33. Dragolea, N.: 9 Drones that will revolutionize Healthcare. Doctor Preneurs, London (2019)
34. D'Andrea, R.: Guest editorial can drones deliver?. IEEE Trans. Autom. Sci. Eng. **11**(3), 647–648 (2014)
35. Gevaers, R., Van de Voorde, E., Vanelslander, T.: Characteristics of innovations in last-mile logistics-using best practices, case studies and making the link with green and sustainable logistics. In: Association for European Transport and Contributors (2009)
36. Murray, C.C., Chu, A.G.: The flying sidekick traveling salesman problem: optimization of drone-assisted parcel delivery. Trans. Res. C Emerging Technol. **54**(7), 86–109 (2015)
37. Jeong, H.Y., Song, B.D., Lee, S.: Truck-drone hybrid delivery routing: payload-energy dependency and No-Fly zones. Int. J. Prod. Econ. **214**, 220–233 (2019). https://doi.org/10.1016/j.ijpe.2019.01.010.
38. Karak, A., Abdelghany, K.: The hybrid vehicle-drone routing problem for pick-up and delivery services. Transp. Res. C Emerging Technol. **102**, 427–449 (2019)
39. Dayarian, I., Savelsbergh, M., Clarke, J.-P.: Same-day delivery with drone resupply. Technical report Milton Stewart School of Industrial and Systems Engineering . . . (2018)
40. Luo, H., Niu, Y., Zhu, M., Hu, X., Ma, H.: Optimization of Pesticide Spraying Tasks via Multi-UAVs Using Genetic Algorithm. Math. Problems Eng. **2017**, 16 (2017)
41. Kim, S.J., Lim, G.J., Cho, J., Côté, M.J.: Drone-aided healthcare services for patients with chronic diseases in rural areas. J. Intell. Rob. Syst. **88**(1), 163–180 (2017)
42. Larsen, A.: The dynamic vehicle routing problem. Citeseer, New York (2000)
43. Applegate, D.L., Bixby, R.E., Chvátal, V., Cook, W.J.: The traveling salesman problem. In: Princeton Series in Applied Mathematics. Princeton University, Princeton (2006)
44. Salhi, S.: Heuristic search: the emerging science of problem solving. Springer, Berlin (2017)
45. Laporte, G., Gendreau, M., Potvin, J.Y., Semet, F.: Classical and modern heuristics for the vehicle routing problem. Int. Tran. Oper. Res. **7**(4–5), 285–300 (2000)
46. Madani, B., Ndiaye, M.: Autonomous Vehicles Delivery Systems Classification: Introducing a TSP With a Moving Depot. In: Proceedings of the 8th International Conference on Modeling, Simulation and Applied Optimization, Bahrain (2019)
47. Asymmetric traveling salesman problem (ATSP). Ruprecht-Karls-Universität Heidelberg (2018). https://www.iwr.uni-heidelberg.de/groups/comopt/software/TSPLIB95/atsp/ (Accessed January 9, 2018)

Chapter 8
Routing Electric Vehicles with Remote Servicing

Rafael Kendy Arakaki, Lucas Porto Maziero, Matheus Diógenes Andrade, Vitor Mitsuo Fukushigue Hama, and Fábio Luiz Usberti

Abstract This paper introduces a problem called electric capacitated covering tour problem (ECCTP), a variant of the vehicle routing problem that allows customers demands to be serviced remotely by electric vehicles with limited battery range and that recharge at alternative fuel stations (AFSs). The ECCTP integrates two research areas: vehicle routing problems and green logistics. We propose a mixed integer linear programming (MILP) mathematical formulation and a biased random-key genetic algorithm (BRKGA) metaheuristic for the ECCTP. A set of benchmark instances from literature is adapted for the problem. Computational experiments show the effectiveness of the proposed methods while providing useful information for the decision-making on transportation operated by electric vehicles.

8.1 Introduction

In the vehicle routing problem (VRP) [1] a set of customers must be serviced by a homogeneous fleet of vehicles with limited capacity, each vehicle starting from a common depot. The objective is to minimize the total cost of vehicle routes. Many variants of the VRP were studied in the literature where additional constraints related to scheduling, budget, fleet heterogeneity, and others are considered [2].

In another research direction, green logistics have emerged as a discipline focused on the planning and operation of logistic systems aiming at high energy efficiency and low levels of carbon emission [3]. Among VRP variants within this discipline, there is the green vehicle routing problem (G-VRP), comprehending additional challenges associated with operating a fleet of alternative fuel vehicles (AFV) and incorporating stops at alternative fuel stations (AFSs) [4]. One of the

R. K. Arakaki (✉) · L. P. Maziero · M. D. Andrade · F. L. Usberti
Institute of Computing, University of Campinas, Campinas, SP, Brazil
e-mail: rafael.arakaki@ic.unicamp.br; lucas.maziero@ic.unicamp.br; fusberti@ic.unicamp.br

V. M. F. Hama
Graduate School of Informatics, Nagoya University, Nagoya, Japan
e-mail: vitorhama@nagoya-u.jp

© Springer Nature Switzerland AG 2020   147
H. Derbel et al. (eds.), *Modeling and Optimization in Green Logistics*,
https://doi.org/10.1007/978-3-030-45308-4_8

biggest challenges modeled by G-VRP is the reduced number of alternative fuel stations and the low range of alternative fuel vehicles. A G-VRP solution example is depicted in Fig. 8.1, in which each customer is visited by a route and some AFSs are visited more than once for recharging.

In another branch of VRP variants, [5] investigated the covering tour VRP, a problem where some customers may be located in regions of difficult access but can be serviced by covering: the vehicle visits one of the neighboring customers instead of directly visiting the customer. This problem has application in determining post-boxes locations within a set of candidate locations, finding optimal collection routes [6]. Another application is to design routes for mobile health care delivery teams where services are rendered at a number of locations by medical teams, and the population living outside these locations must travel on foot to reach them [7, 8].

This work merges two branches of vehicle routing problems, the G-VRP and the covering tour VRP, resulting in a new problem called electric capacitated covering tour problem (ECCTP). In the ECCTP, a homogeneous fleet of electric vehicles are used to service the demands of customers. A customer can be serviced by any vehicle visiting a node inside the neighborhood of that customer. An empty neighborhood implies that the corresponding customer must be visited in order to be serviced. Figure 8.2 presents an ECCTP example in which some vertices are customers and others are traffic vertices. A traffic vertex does not have demand and may be visited or not. Both traffic and customer vertices can be used to service demands of neighboring customers by covering.

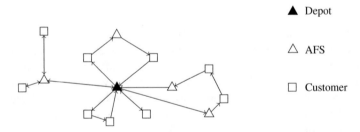

Fig. 8.1 G-VRP solution example

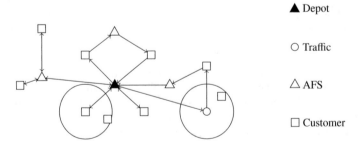

Fig. 8.2 ECCTP solution example

Our Contribution We introduce a new VRP variant called electric capacitated covering tour problem (ECCTP) and present a mixed integer linear programming (MILP) model for the problem. A biased random-key genetic algorithm (BRKGA) is also proposed for the ECCTP. Furthermore, a set of benchmark instances for the problem was created based on public benchmark instances originally proposed for the VRP. Computational experiments were conducted with the proposed methods to evaluate the effectiveness of the approaches and to extract useful information for transportation decision-making.

Section 8.2 presents a literature review of papers related to this work, classified in two subsections: (1) covering tour problems and (2) green vehicle routing problems. In Sect. 8.3 the ECCTP is formally defined. In Sect. 8.4 a mathematical formulation for the problem is presented. Section 8.5 describes the BRKGA metaheuristic. Section 8.6 contains the results and analysis of the computational experiments. Finally, Sect. 8.7 presents the final remarks.

8.2   Literature Review

8.2.1   Covering Routing Problems

The covering salesman problem (CSP), proposed by J. R. Current and D. A. Schilling [9], addresses the following question: considering a set of sites scattered in the plane that must be covered by a single vehicle tour and knowing that each site covers some of its neighbors, what is the minimum length of an enclosed vehicle tour in which all sites are covered? More formally, given an undirected graph, the CSP objective is to find the shortest Hamiltonian cycle on a subset of vertices that covers the graph. The special case where each vertex covers strictly itself is the traveling salesman problem (TSP) [10], which follows that CSP is also NP-hard.

Some solution methodologies were proposed in the literature for the CSP. J. R. Current and D. A. Schilling [9], for example, developed a two-step heuristic to solve the CSP: the first step solves a set cover problem; the second step solves the TSP on the vertices determined by the first step. More than two decades later [11] revisited the problem by proposing a heuristic for the CSP embedded within an integer linear programming (ILP) framework. First they employ constructive heuristics to find good initial solutions and then the tour vertices are rearranged by the use of ILP techniques in an attempt to reduce its length. More recently, [12] give a polynomial size formulation and a hybrid heuristic for the CSP. Their heuristic combines ant colony optimization and dynamic programming.

M. Gendreau et al. [13] studied the covering tour problem (CTP), a variant of CSP. Let $G = (V \cup W, E)$ be an undirected graph, where $V \cup W$ is the set of vertices and E is the set of edges. Vertex v_0 is the depot, V is the set of vertices that can be visited, $T \subseteq V$ is the set of vertices that must be visited ($v_0 \in T$), and W is the set of vertices that must be covered but cannot be visited. The goal of the CTP is to determine a minimum length tour that visits a subset of vertices $S \subseteq V$ such that $T \subseteq S$ and each vertex of W is covered by some vertex in S. The authors proposed heuristics and a branch-and-cut algorithm to solve the CTP.

There are works in the literature that tackle the geometric version of CSP. In this version, a compact region of the plane containing each vertex is specified as a neighborhood set. The goal is to find a minimum length tour that starts from a depot and intercepts all neighborhood sets, thus covering all its corresponding vertices. Approximation algorithms, heuristics, and methodologies based on ILP were developed for this version [14–17].

Some works in the literature address variants of CTP considering multiple vehicles. [18] introduced the multi-vehicle covering tour problem (m-CTP). Given a graph $G = (V \cup W, E)$, where $V \cup W$ is the set of vertices and E the set of edges. Vertex v_0 is a depot at which m identical vehicles start their routes, V is the set of vertices that can be visited, $T \subseteq V$ is the set of vertices that must be visited ($v_0 \in T$), and W is the set of vertices that must be covered but cannot be visited. The objective of m-CTP is to determine m routes of minimum total length satisfying the following restrictions:

- there are at most m vehicle routes and each route starts and ends at vertex v_0;
- each vertex of T is visited exactly once, while each vertex of $V \backslash T$ is visited at most once;
- each vertex of W must be covered by a route in the sense that it must lie within a preset distance c of a vertex of V belonging to a route (assuming that v_0 does not cover all vertices of W);
- the number of vertices (excluding v_0) in each route is limited by a value p;
- the length of each route cannot exceed a value q.

The work of [19] investigated the location of distribution centers in the context of disaster relief. In these situations, support teams are unable to visit each affected area. Therefore, people need to go to a particular distribution center to obtain survival items, provided that these centers are not too far away. The locations of the distribution centers are defined for a car fleet departing from a depot. This application can be modeled through m-CTP. The authors proposed an integer linear programming (ILP) formulation and developed a heuristic able to produce high-quality solutions, even for large size instances.

Some works in the literature considered capacitated versions of covering vehicle routing problems. S. Allahyari et al. [5] proposed the multi-depot covering vehicle routing problem (MDCTVRP). Let $G = (N, A)$ be a directed graph, where $N = N_c \cup N_d$ is the set of vertices; $A = \{(i, j) | i, j \in N\}$ is the set of arcs; $N_c = \{1, 2, ..., n_c\}$ is the set of customers' vertices such that all $i \in N_c$ has a demand $d_i > 0$; $N_d = \{1, 2, ..., n_d\}$ is the set of depots of which the vehicles begin their routes. In the MDCTVRP it is not necessary that each customer has to be visited by a vehicle, provided that they are within a maximum distance from at least one visited customer. The authors developed two formulations of mixed integer linear programming (MILP) and a greedy randomized adaptive search procedure (GRASP) metaheuristic to solve the MDCTVRP.

In the work of [20] the multi-vehicle cumulative covering tour problem (m-CCTP) was proposed. Given a complete graph $G = (V \cup W, E)$, where $V = \{v_0, ..., v_{+}n\}$ is the set of vertices. $V' = V \backslash \{v_0, v_{n+1}\}$ is the set of vertices that can be visited and v_0 and v_{n+1} are the depots; W is the set of inaccessible vertices that must be covered; and E is the set of edges. Let $T \subset V'$ be the set of vertices that must be visited. The m-CCTP consists of finding a set of routes that visits all customers in T and some in $V' \backslash T$ such that the vertices in W are covered and the sum of arrival times at the visited vertices is minimized. An MILP formulation and a GRASP metaheuristic are proposed for m-CCTP. The computational experiments evaluated the performance of the methodologies for a set of instances and showed the effectiveness of the GRASP metaheuristic.

8.2.2   Green Vehicle Routing Problems

S. Erdoğan and E. Miller-Hooks [4] proposed the G-VRP and developed two constructive heuristics: the modified heuristic of Clarke and Wright (savings) and the algorithm of clustering based on density. The authors also proposed a heuristic custom improvement technique.

Ç. Koç and I. Karaoglan [21] proposed an approach for the G-VRP that utilizes a simulated annealing within a branch-and-cut framework. The simulated annealing was used to improve the initial solution and to find better primal bounds during the search. The authors evaluated their approach in terms of the number of optimal solutions obtained and the computational time required to find the best solution. The computational results have shown that 22 of 40 instances with 20 customers were solved optimally with a pre-defined time limit. Moreover, [21] proposed a new mathematical formulation with fewer variables and restrictions than previous works and without the need for network augmenting.

V. Leggieri and M. Haouari [3] proposed an improved model, achieving better results than the previous model by Ç. Koç and I. Karaoglan [21] and S. Erdoğan and E. Miller-Hooks [4]. Their formulation offers two significant advantages: compactness, due to a polynomial number of variables and restrictions, and flexibility, allowing it to be more easily adapted to other VRP variants. The main idea was to apply reformulation-linearization technique (RLT), proposed by H. Sherali and W. Adams [22] and H. D. Sherali and W. P. Adams [23]. First a non-linear MILP formulation is proposed for the G-VRP, and then an equivalent MILP formulation is derived. V. Leggieri and M. Haouari [3] provide empirical evidence that the formulation and the reduction procedure was able to find optimal solutions for medium-sized instances using a general purpose solver.

A. Montoya et al. [24] propose a two-phase heuristic for the G-VRP. In the first phase, the heuristic constructs a set of routes using route-first cluster-second randomized heuristics with a procedure of optimal insertion of AFSs. In the second phase, a G-VRP solution is obtained from the resolution of set partitioning formulation over the set of routes constructed in the first phase. To test their approach, [24] execute experiments in a set of 52 literature instances. The results showed that the heuristic is competitive with other state-of-the-art methods.

Some very similar problems related to the G-VRP are proposed by R. G. Conrad and M. A. Figliozzi [25] and Y.-W. Wang et al. [26]. For example, they introduced the recharging vehicle routing problem, where vehicles with limited range are allowed to recharge at customer locations mid-tour. Their work addresses the problem in two versions: the capacitated (CRVRP) and the capacitated with time-windows (CRVRP-TW), for both presenting mathematical formulations and experimental results with solution bounds.

Another variant which connects routing problem to green logistics is the pollution-routing Problem (PRP), introduced by T. Bektaş and G. Laporte [27]. The PRP has a broader and more comprehensive objective function that accounts not just for the travel distance, but also for the number of greenhouse emissions, fuel, travel times, and their costs. The results of [27] suggest that the PRP has the potential of yielding savings in total cost.

8.3   Problem Description

The ECCTP is defined next. Consider a complete graph $G(V, E)$, where V is the set of vertices and E is the set of edges. The set of vertices $V = W \cup T \cup F \cup \{v_0\}$ is composed of four disjoint sets of vertices: W contains customer vertices; T contains traffic vertices; F contains alternative fuel stations (AFFs) vertices; and v_0 is the depot vertex. The edges $(i, j) \in E$ have costs $c_{ij} \geq 0$. There is a set of M_{UB} homogeneous vehicles that start their routes in the depot vertex v_0 with a demand capacity Q and a battery level charged to the maximum of β units. Each customer $w \in W$ has an associated demand $d_w \geq 0$. For each customer $w \in W$, the set $C(w)$ is the subset of all vertices in $W \cup T$ that cover w. It is considered that $w \in C(w), \forall w \in W$. The vehicles must service all customers in W. The goal of ECCTP consists of finding a set of $M \leq M_{UB}$ minimum cost routes satisfying the following constraints:

- Each route begins and ends at the vertex depot v_0.
- Each vertex $v \in (T \cup W)$ is visited at most once by only one route.
- Each vertex $f \in F$ can be visited multiple times by several routes.
- Each vertex $w \in W$ must be covered by at least one route, i.e., it must be serviced through some vertex $v \in C(w)$ that is visited by some route.
- The demand d_w of a vertex $w \in W$ is serviced exclusively by one route.
- The demand of a vertex $w \in W$ can be serviced by a route $k \in \{1, \ldots, M_{UB}\}$ only if there is a vertex v visited by route k such that $v \in C(w)$.
- The total demand serviced by a route must not exceed the capacity Q of the vehicle.
- The battery level of a vehicle decreases by c_{ij} when it travels along an edge $(i, j) \in E$. It is not allowed for a vehicle to visit an edge whose cost is higher

than its current battery level. When the vehicle visits a station vertex $f \in F$, its battery level is restored to the maximum capacity of β units.

The VRP is a special case of the ECCTP where: (1) only depot and customer vertices are considered; (2) a very high battery capacity ($\beta = \infty$); (3) each customer covers only itself ($\forall w \in W : C(w) = \{w\}$). Since the VRP is a NP-hard problem [1], it follows that the ECCTP is also NP-hard.

8.4 Mathematical Model

The ECCTP formulation is defined on a subgraph $G'(V, E')$ of the original graph $G(V, E)$. The subset of edges $E' = E \setminus E^{inf}$ is the original set E without some infeasible edges, where $E^{inf} = \{(v_i, v_j) \in E : c_{ij} > \beta\}$ is a set of edges that cannot be visited by any feasible solution because of the limited battery capacity.

The functions and parameters used in the MILP formulation for the ECCTP are:

- $\delta^+(i)$: is the set of directed edges in E' which *leave* a vertex $v_i \in V$.
- $\delta^-(i)$: is the set of directed edges in E' which *enter* a vertex $v_i \in V$.
- $\delta(i) = \delta^+(i) \cup \delta^-(i)$.
- $M_{set} = \{1, ..., M_{UB}\}$ is the set of available vehicles.
- $M_{LB} = \lceil \sum_{i \in W} d_i / Q \rceil$ is a lower bound for a feasible number of vehicles.

The sets of variables used in the MILP formulation for the ECCTP are given next:

- x_{ij}^k: is the number of times edge $(v_i, v_j) \in E'$ is visited by route $k \in M_{set}$.
- y_i : is the number of times vertex $v_i \in (W \cup T)$ is visited.
- w_f : is the number of times vertex $v_f \in F$ is visited.
- z_i^k : is 1 if vertex $v_i \in W$ is served by route $k \in M_{set}$; 0 otherwise.
- e_i : is the vehicle battery level at arriving in vertex $v_i \in V$.
- M : is the number of routes.

The objective function (8.1) minimizes the routes total cost. Constraints (8.2), (8.3), and (8.4) imply that the depot must belong to all routes and limit the number of routes. Constraints (8.5) guarantee that every vertex $v_i \in W$ has its demand serviced. Constraints (8.6) ensure that for each route the number of edges that enters and leaves a vertex is the same. Constraints (8.7) and (8.8) integrate edge variables with vertex degree variables of customer/traffic vertices and stations, respectively. Constraints (8.12) ensure that, for every vertex $v_w \in W$ whose demand is attended by vehicle $k \in M_{set}$, there is at least one vertex $i \in C(w)$ visited by vehicle k, where $C(w)$ is the set of vertices that cover v_w. Constraints (8.13) ensure that the total attended demand by a vehicle must not exceed its capacity Q. Constraints (8.14) are the connectivity constraints which assure that all routes are connected to the depot. More specifically, it states that given a set of vertices S without depot vertex v_0 and a customer/traffic vertex $v_{i*} \in S$, if vertex v_{i*} is visited by route k then there must be at least one edge entering the set S in that route in order to v_{i*} be connected to the depot.

$$\text{Min} \sum_{(i,j)\in E'} c_{ij} x_{ij} \tag{8.1}$$

s.t.

$$\sum_{k\in M_{set}} \sum_{j\in\delta^+(v_0)} x_{ij}^k = \sum_{k\in M_{set}} \sum_{j\in\delta^-(v_0)} x_{ji}^k = M \tag{8.2}$$

$$M_{LB} \leqslant M \leqslant M_{UB} \tag{8.3}$$

$$\sum_{j\in\delta^+(v_0)} x_{ij}^k \leqslant 1 \qquad \forall k \in M_{set} \tag{8.4}$$

$$\sum_{k\in M} z_i^k = 1 \qquad \forall v_i \in W \tag{8.5}$$

$$\sum_{j\in\delta^+(i)} x_{ij}^k = \sum_{j\in\delta^-(i)} x_{ji}^k \qquad \forall v_i \in V, \forall k \in M_{set} \tag{8.6}$$

$$\sum_{k\in M_{set}} \sum_{j\in\delta^+(i)} x_{ij}^k = y_i \qquad \forall v_i \in W \cup T \tag{8.7}$$

$$\sum_{k\in M_{set}} \sum_{j\in\delta^+(f)} x_{ij}^k = w_f \qquad \forall v_f \in F \tag{8.8}$$

$$\sum_{i\in C(w)} \sum_{j\in\delta^-(i)} x_{ji}^k \geqslant z_w^k \qquad \forall v_w \in W, k \in M_{set} \tag{8.9}$$

$$\sum_{i\in W} d_i z_i^k \leqslant Q \qquad \forall k \in M \tag{8.10}$$

$$\sum_{v_i\in(V\backslash S)} \sum_{v_j\in S} x_{ij}^k \geqslant \sum_{v_p\in\delta^-(v_{i*})} x_{pi*}^k \qquad \forall S \subseteq V \backslash \{v_0\}$$

$$\forall v_{i*} \in S \cap (W \cup T), \forall k \in M_{set} \tag{8.11}$$

$$\sum_{i\in C(w)} \sum_{j\in\delta^-(i)} x_{ji}^k \geqslant z_w^k \qquad \forall v_w \in W, k \in M_{set} \tag{8.12}$$

$$\sum_{i\in W} d_i z_i^k \leqslant Q \qquad \forall k \in M \tag{8.13}$$

$$\sum_{v_i\in(V\backslash S)} \sum_{v_j\in S} x_{ij}^k \geqslant \sum_{v_p\in\delta^-(v_{i*})} x_{pi*}^k, \forall S \subseteq V \backslash \{v_0\} \quad \forall v_{i*} \in S \cap (W \cup T), \forall k \in M_{set}$$

$$\tag{8.14}$$

$$0 \leqslant e_i \leqslant \beta \qquad \forall v_i \in V \cup W \qquad (8.15)$$

$$e_f = \beta \qquad \forall v_f \in F \cup \{v_0\} \qquad (8.16)$$

$$e_j \leqslant e_i - c_{ij} \left(\sum_{k \in M_{set}} x_{ij}^k \right) + \beta \left(1 - \left(\sum_{k \in M_{set}} x_{ij}^k \right) \right) \qquad \forall v_j \in W \cup T, \forall v_i \in V$$

$$(8.17)$$

$$c_{ij} \left(\sum_{k \in M_{set}} x_{ij}^k \right) \leqslant e_i \qquad \forall v_j \in F \cup \{v_0\}, \forall v_i \in W \cup T$$

$$(8.18)$$

$$x_{ij}^k \in \{0, 1\} \qquad \forall (v_i, v_j) \in E' : v_i \in W \cup T \text{ or } v_j \in W \cup T \qquad \forall k \in M_{set}$$

$$(8.19)$$

$$x_{ij}^k \in \mathbb{Z}^+ \qquad \forall (v_i, v_j) \in E' : v_i, v_j \in F \cup \{v_0\},$$

$$\forall k \in M_{set} \qquad (8.20)$$

$$y_i \in \{0, 1\} \qquad \forall v_i \in W \cup T \qquad (8.21)$$

$$w_f \in \mathbb{Z}^+ \qquad \forall v_f \in F \qquad (8.22)$$

$$z_i^k \in \{0, 1\} \qquad \forall v_i \in W, \forall k \in M_{set} \qquad (8.23)$$

$$e_j \in \mathbb{R}^+ \qquad \forall v_j \in V \qquad (8.24)$$

$$M \in \mathbb{Z}^+ \qquad (8.25)$$

Constraints (8.15)–(8.18) enforce that the vehicles can only visit vertices that are within the reach of their residual capacities. Constraints (8.15) define the minimum and maximum energy level. Constraints (8.16) restore the vehicle energy level when a vehicle is visiting a depot or a recharging station. Constraints (8.17) update the vehicle energy level when a vertex $j \in W \cup T$ is visited. Constraints (8.18) do not allow vehicles to recharge in stations or return to depot without the required energy. Constraints (8.19)–(8.25) define the domain for each set of variables.

There are an exponential number of connectivity constraints (8.14); therefore, a complete enumeration of them is only possible for very small instances. The solution adopted was to consider an iterative algorithm that adds the connectivity constraints to the formulation as they are needed to progress the optimization. First the formulation is executed without these constraints and as soon as an integer solution (x^*, M) is obtained, the iterative algorithm is called. The process is shown in Algorithm 1. A connected component algorithm is executed M times for each route $k \in \{1, ..., M\}$ on undirected graphs $G_{x^*}^k (V_{x^*}^k, E_{x^*}^k)$ induced by the edges visited in each route k of the solution. From the connected components in each induced graph one can observe that excluding the one that contains the depot vertex v_0, all others make subset of vertices disconnected from the depot. Therefore, for each subset of vertices S associated with a component disconnected from v_0, a set of corresponding connectivity constraints (8.14) is added to the formulation, then the *solver* restarts the optimization. This process is repeated until an optimal solution that does not violate any of the connectivity constraints (8.14) is obtained.

Algorithm 1: Iterative algorithm for ECCTP formulation

Data: A feasible solution (x^*, M) for constraints (2-10) and (12-22).
Result: A set CC of connectivity constraints (11) violated by (x^*, M), if there is any.
for $k = 1$ *to* M **do**
  Create a graph $G^k_{x^*}(V^k_{x^*}, E^k_{x^*})$ induced by the set of edges
  $E^k_{x^*} = \{(i, j) \in E' : \sum_{j \in \delta^+(v_i)} x^k_{ij} + \sum_{j \in \delta^-(v_i)} x^k_{ji} \geqslant 1\}$ and where
  $V_{x^*} = \{i \in V : \sum_{j \in \delta^+(v_i)} x^k_{ij} \geqslant 1\}$;
  Obtain the set of connected components in $G^k_{x^*}$;
  for each found connected component S such that $v_0 \notin S$ **do**
    $v_{i^*} \leftarrow$ an arbitrarily chosen vertex from $S \cap (W \cup T)$;
    $CC_{new} \leftarrow$ set of connectivity constraints (11) given by (S, v_{i^*}) for each $k \in M_{set}$;
    $CC \leftarrow CC \cup CC_{new}$;

return CC ;

One of the advantages of our proposed formulation for ECCTP, comparing to other formulations for similar electric vehicle routing problems in literature, is: (1) the ability to handle any number of visits on each station vertex; (2) allowing solutions that have any number of consecutive visits of stations. For example, [4] proposed a formulation where dummy vertices are created for each possible visit to a station; [4, 21] and [3] proposed formulations that assume a vehicle never makes two consecutive visits in station vertices; and [26] proposed a formulation that assumed each station would be used at most once per vehicle.

8.5   BRKGA

In this section, we will describe a biased random-key genetic algorithm (BRKGA) metaheuristic developed for the ECCTP. The BRKGA [28] is a metaheuristic proposed to address combinatorial optimization problems. The key idea is developing a *decoder* function to map a sequence of fixed-length real-valued numbers (*random keys*) to represent a solution for the targeted problem. This section first gives an overview of this metaheuristic and then our method is described.

8.5.1   Biased Random-Key Genetic Algorithm (BRKGA)

Similar to genetic algorithms, BRKGAs represent solutions by *chromosomes*. The method starts with an initial population of random *individuals*. The population is composed of two ranks: elite and non-elite individuals. The algorithm then processes the generations iteratively in four steps: (1) *crossover* of couples made by an elite and a non-elite individual; (2) small fraction of pure random mutants are created; (3) decoding and ranking the new generation; and (4) a population selection procedure to keep the population at a fixed size.

The main characteristics that define the BRKGA, in contrast to the general class of genetic algorithm, are that in the BRKGA: (1) the solutions are represented by a fixed-length sequence of N random numbers, each in the interval $[0, 1)$; (2) elitist strategy is always adopted: best solutions pass to next generation without change; (3) the *crossover* is always between an elite and a non-elite individual; and (4) no *mutation* method is considered: instead, some random individuals (*mutants*) are inserted in the population from time to time.

The BRKGA contains only one problem-specific procedure: a *decoder* function. This procedure starts from a vector of *random keys*, i.e., real-valued numbers in the interval of $[0, 1)$ and should output a solution for the problem that can be evaluated by a *fitness* function. Therefore, to define a BRKGA method, one just needs to define a corresponding decoding function and its problem specified parameters [29]. The parameters chosen for the BRKGA are described in Sect. 8.6. In the following subsection, the proposed decoder function for ECCTP is described.

8.5.2 Decoding an ECCTP Solution from a Vector of Random Keys

The decoder function for ECCTP is composed of the following steps:

1. construction of a single route with unlimited capacity and an unlimited electric battery that visit exactly once each customer and traffic vertex in $(W \cup T)$;
2. assign which customers will be served by covering each vertex in the route;
3. Split procedure: optimally split the single route with unlimited capacity into a set of routes with feasible capacity, each starting and ending in the depot vertex v_0.
4. InsertStations procedure: a dynamic programming approach to optimally insert the stations' visits in each route, making the route feasible.

In the first step, a chromosome S of length $N = |W \cup T|$ is given. The alleles in non-decreasing order keeping track of the original positions as shown in Fig. 8.3. Each position is then mapped to a unique element in $W \cup T$. The result is a single route R that visits each vertex in $W \cup T$ once as shown in Fig. 8.4. This route may have a demand greater than the limited capacity and/or may use more energy than the battery supports. These infeasibilities will be treated by the following steps.

In the second step, we assign to each vertex in R a subset of customers that will be serviced (by covering) when the route visits that vertex. Stating from the vertex r_1 at the beginning of the route $R = (r_1, r_2, \ldots, r_n)$, we assign to node r_i the following customers, in that order: (1) r_i itself if it is a customer vertex not yet assigned; (2)

Chromossome S	0.4	0.5	0.1	0.8	0.3	0.9
	1	2	3	4	5	6

Chromossome S'	0.1	0.3	0.4	0.5	0.8	0.9
Route R	3	5	1	2	4	6

Fig. 8.3 Decoding a chromosome into a route

Fig. 8.4 A single route decoded from the chromosome

the customers, in lexicographic order of, that can be served by covering from r_i and which have not yet been assigned, until the total assigned demand does not exceed the vehicle capacity. If all customers coverable by r_i were assigned and there is still capacity remaining, we try to assign customers to the next vertex r_{i+1} and so on. If the demand was greater or equal to the capacity, we stop assigning customers to the vertex, reset the total demand, and restart assigning customers to the following vertices of the route. In the example shown in Fig. 8.5, vertex 3 services customers 3, 2, and 4, while vertices 5 and 1 only services themselves, and vertex 4 only serves customer 6. Vertices 2 and 6 do not service any client and will be removed in the next step. This guarantees that every customer will be serviced exactly once, but still does not assure the battery constraint.

Next, we remove from R all the vertices that have not been assigned to service any customer. Then the split procedure is applied in order to handle capacity constraints. This procedure was inspired by the split procedure originally proposed for VRP [30]. The single route R is divided into several sub-routes such that each one attends the capacity constraints. This procedure is optimal for a given sequence of vertices. We create a weighted direct acyclic graph (DAG) whose vertices are ordered as same as the route R. Edge (i, j) exists if and only if the sub-route $0 \rightarrow i \xrightarrow{\text{vertices between } i,j} 0$ (where 0 represents the depot), has a demand less or equal to the vehicle capacity Q. The weight of this edge is the total traveled distance of this route. By solving the shortest path problem in this DAG we can find the routes (edges) such that the total distance is minimum. An example is illustrated in Fig. 8.6, where the obtained shortest path cost is $2 + 5 = 7$. This algorithm is fast since the time complexity for the shortest path in DAGs is $O(|E|)$.

In the last step, InsertStations procedure, we solve the problem of deciding where the vehicle should proceed to recharge its battery for each route. The pseudocode of

Customer/Transit node	Can cover		
1	1	5	3
2	2	1	4
3	3	2	4
4	4	2	3
5	5	1	2
6	5	4	3

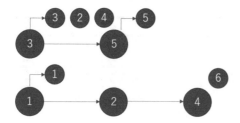

Fig. 8.5 Assign customers to a vertex

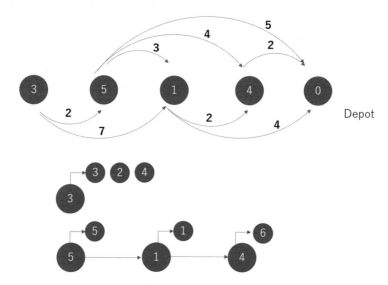

Fig. 8.6 The split procedure

the dynamic programming is shown in Algorithm 2. One can observe that between two consecutive vertices i, j of a given route, we have two choices: go from i to a recharge station not farther than the current battery level and then visit j (line 2); or go straight from i to j (line 2). In the first case, if we have more than one feasible ways of going from a station to j, with the same remaining battery level, then the shortest way is chosen and saved in the table (lines 2 and 2). An example is

Autonomy $\beta = 5$

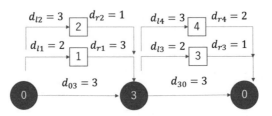

	0	1	2	3	4	5
0	∞	∞	∞	∞	∞	0
3	∞	∞	$\min\{d_{03}, d_{l1} + d_{r1}\}$ $= d_{03} = 3$	∞	$d_{l2} + d_{r2}$ $= 4$	∞
0	∞	∞	∞	$4 + d_{l4} + d_{r4}$ $= 9$	$\min\{3 + d_{l3} + d_{r3}, 4 + d_{l3} + d_{r3}\}$ $= 6$	∞

Fig. 8.7 Inserting recharging stations

illustrated in Fig. 8.7, where the chosen decision was to go from vertex 0 to vertex 3 directly and then from vertex 3 to recharge in station 3 and then arrive at vertex 0. The total route cost is 6. After InsertStations procedure is executed for each route, the obtained solution is feasible for ECCTP.

Algorithm 2: InsertStations

Data: A route R, vehicle battery range β, shortest distance between any two vertices $d(i, j)$.
Result: A feasible route.
Create a matrix $M[1, \ldots, |R|][0, \ldots, \beta]$;
Initialize every value of M to ∞;
Initialize $M[1][\beta] := 0$;
for $i = 1$ *to* $|R| - 1$ **do**
  for $j = 0$ *to* β **do**
    if $M[i][j]! = \infty$ **then**
      $dist = d(i, i + 1)$;
      for each station k such that $d(i, k) <= j$ and $d(k, i + 1) <= \beta$ **do**
        $d_l = d(i, k)$;
        $d_r = d(k, i + 1)$;
        if $M[i][j] + d_l + d_r < M[i + 1][\beta - d_r])$ **then**
          $M[i + 1][\beta - d_r] = M[i][j] + d_l + d_r$;
      if $j\text{-}dist >= 0$ **then**
        if $M[i][j] + dist < M[i + 1][j - dist]$ **then**
          $M[i + 1][j - dist] = M[i][j] + dist$;

return *The route corresponding to the* $\min_j M[|R|][j]$;

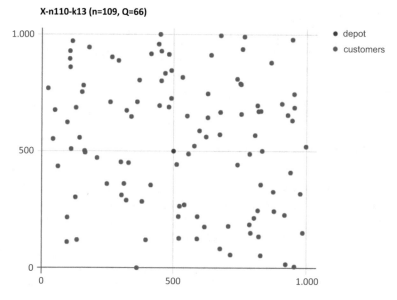

Fig. 8.8 The CVRP instance X-$n110$-$k13$ used to generate the instances for ECCTP[1]

8.6   Computational Experiments

8.6.1   Benchmark of Instances

The set of instances for the ECCTP was generated from a set of CVRP instances proposed by E. Uchoa et al. [31], namely X-$n110$-$k13$. This instance is named in the format X-nA-kB, where A represents the $n + 1$ vertices (including the depot vertex) and B the minimum number of vehicles needed to serve the demand for all vertices, calculated by solving the bin packing problem. Figure 8.8 illustrates the CVRP instance X-$n110$-$k13$ (Table 8.1).

From that X-$n110$-$k13$ instance, a total of 54 ECCTP instances were generated with varying parameters. Table 8.1 summarizes the parameters used in the generation of instances for the ECCTP. Each generated ECCTP instance considered the following parameters:

In the ECCTP instances the depot vertex is the same as the one in X-$n110$-$k13$ and the remaining vertices are the $|V| - 1$ closest vertices to the depot. The vehicle capacity Q of the ECCTP instances is the same as the X-$n110$-$k13$ instance.

The customer and station vertices were chosen randomly, with uniform distribution, from the set $V \setminus \{v_0\}$. The $|V| - 1 - |W| - |F|$ remaining vertices are traffic

---

[1]Image produced by E. Uchoa et al. [31] and extracted from the site http://vrp.atd-lab.inf.puc-rio. br/index.php/en/plotted-instances?data=X-n110-k13

Table 8.1 Parameters of generated instances for the ECCTP

Parameter	Description	Values
$|V|$	Total of vertices	$\{21, 31, 41, 51, 61, 71\}$
$|W|$	Number of customer vertices	50% of $|V| - 1$
$|T|$	Number of traffic vertices	30% of $|V| - 1$
$|F|$	Number of station vertices	20% of $|V| - 1$
β	Vehicle battery capacity	$\{50\%, 60\%, 70\%\}$ of d_G
c	Size of coverage	$\{5, 7, 11\}$

d_G—graph diameter of the ECCTP instance

vertices. The customer vertices are always within a maximum distance of $\frac{\beta}{2}$ from some station vertex. Besides, it is guaranteed there exists a tree spanning all stations and the depot using only edges of length β or less. Therefore, the vehicles can travel through all customer and station vertices can be served.

For each vertex $v \in (W \cup T) \setminus \{v_0\}$ a set $D(v)$, which represents the vertices covered by v, is formed by the c closest customers or traffic vertices to v. From that, the set $C(w)$, which represents the vertices that cover a customer $w \in W$, is generated by simply considering that $v \in C(w) \iff w \in D(v)$.

The instances for the ECCTP are named in the format $wI\text{-}eaJ\text{-}cK$, where I represents the number of customer vertices, J the vehicle battery capacity, and K the size of coverage.

8.6.2 Computational Settings

The computational experiments were executed in an Intel Xeon E3-1230 V2 3.3 GHz with 32 GB of RAM and Linux 64-bit operating system. The MILP formulation was implemented using solver Gurobi Optimizer 8.1.1, set with a 3600 s time limit for all instances. The BRKGA described in this paper was implemented using the C++ framework proposed by R. F. Toso and M. G. C. Resende [29], set with a $|W| * 10$ s time limit for each instance. Also, the parameters used by BRKGA are:

- Number of alleles per chromosome: $|W| + |T|$;
- Number of chromosomes in population: 100;
- Size of the elite set in population: 10% of the entire population;
- Number of mutants to be introduced in the population at each generation: 10% of the entire population;
- Probability that an allele is inherited from the elite parent: 80%.

8.6.3 Experiment Results

The results of the computational experiments for the benchmark of instances are reported in Tables 8.2, 8.3, and 8.4.

Table 8.2 Summary results of experiments

	MILP formulation			BRKGA		
$|W|$	Avg gap (%)	# Opt	Avg time	Avg gap (%)	# Opt	Avg time
10	0.00	9	1.69	3.42	2	0.02
15	2.60	6	1433.40	6.09	1	6.26
20	13.01	1	3289.03	15.40	0	20.64
25	44.59	0	3600.00	44.63	0	24.33
30	52.69	0	3600.00	46.63	0	65.80
35	85.42	0	3600.00	59.99	0	101.65
Overall	33.05	16	2587.35	29.36	3	36.45

Table 8.2 reports the summary results of the experiments and for each $|W|$ the following data are shown:

- avg gap (%): average optimality gap;
- # opt: number of optimal solutions;
- avg time: average execution time in seconds.

The column group "MILP formulation" reports the summary results obtained by the MILP formulation, while the column group "BRKGA" reports the summary results obtained by the BRKGA metaheuristic. Note that the "avg time" column of the "MILP formulation" column group is the average total execution time of the MILP formulation, while the "avg time" column of the "BRKGA" column group represents the average execution time until the best upper bounds were obtained. The last row reports the average values of each column.

From Table 8.2 one can observe that for the smaller instances ($|W| \leqslant 20$) the solution quality of the MILP formulation was superior, achieving a total of 16 optimal instances, while the BRKGA obtained 3. Conversely, BRKGA has a processing time much faster than the MILP formulation for instances of all sizes: in overall average, the BRKGA is approximately 70 times faster.

Regarding the hardest instances ($|W| \geqslant 30$) the performance of BRKGA is superior in both solution cost and processing time. For $|W| = 35$, for example, BRKGA obtained an average gap (%) of 59.99 in comparison to 85.42 of the MILP formulation. Moreover, for these instances, the BRKGA average time was approximately 35 times faster. This shows that the BRKGA is the most suitable method for the hardest ECCTP instances, with the advantage of also being faster. Conversely, the MILP formulation can be used to obtain good solutions for small instances and to obtain lower bounds.

Tables 8.3 and 8.4 report the full results of the experiments and for each instance, the following data are shown:

- UB: best upper bound obtained;
- LB: best lower bound obtained;
- gap (%): optimality gap ($\frac{UB-LB}{UB}$) $* 100$;
- time: execution time in seconds;
- M: number of vehicles used in the solution.

Table 8.3 Results for instances with $10 \leqslant |W| \leqslant 20$

$|W|$	β	c	M_{UB}	MILP formulation					BRKGA			
				UB	LB	gap (%)	time	M	UB	gap (%)	Time	M
10	50%	5	4	911	911	0.00	0.90	2	911	0.00	0.01	2
		7	4	711	711	0.00	0.80	2	715	0.56	0.02	2
		11	4	278	278	0.00	0.10	2	294	5.44	0.04	2
		Overall		633.33	633.33	0.00	0.60	2.00	640.00	2.00	0.02	2.00
	60%	5	4	819	819	0.00	3.70	2	819	0.00	0.01	2
		7	4	669	669	0.00	1.60	2	688	2.76	0.02	2
		11	4	278	278	0.00	0.10	2	294	5.44	0.02	2
		Overall		588.67	588.67	0.00	1.80	2.00	600.33	2.73	0.02	2.00
	70%	5	4	768	768	0.00	50.00	3	819	6.23	0.01	2
		7	4	654	654	0.00	2.90	2	688	4.94	0.03	2
		11	4	278	278	0.00	0.10	2	294	5.44	0.01	2
		Overall		566.67	566.67	0.00	2.67	2.33	600.33	5.54	0.02	2.00
	Overall			596.22	596.22	0.00	1.69	2.11	613.56	3.42	0.02	2.00
15	50%	5	4	1129	1085	3.90	3600.00	2	1160	6.47	44.47	2
		7	4	942	942	0.00	1070.40	2	948	0.63	7.54	2
		11	4	587	587	0.00	6.10	2	634	7.41	0.10	2
		Overall		886.00	871.33	1.30	1558.83	2.00	914.00	4.84	17.37	2.00
	60%	5	4	1046	938	10.33	3600.00	2	1076	12.83	0.23	2
		7	4	894	894	0.00	707.00	2	926	3.46	0.44	3
		11	4	587	587	0.00	6.70	2	634	7.41	0.04	2
		Overall		842.33	806.33	3.44	1437.90	2.00	878.67	7.90	0.24	2.33
	70%	5	4	1013	920	9.18	3600.00	2	1013	9.18	3.31	2
		7	4	833	833	0.00	283.10	2	833	0.00	0.01	2
		11	4	587	587	0.00	27.30	2	634	7.41	0.19	2
		Overall		811.00	780.00	3.06	1303.47	2.00	826.67	5.53	1.17	2.00
	Overall			846.44	819.22	2.60	1433.40	2.00	873.11	6.09	6.26	2.11
20	50%	5	5	1657	1321	20.28	3600.00	3	1657	20.28	8.87	3
		7	5	1285	1184	7.86	3600.00	3	1366	13.32	149.82	3
		11	5	1061	955	9.99	3600.00	3	1068	10.58	0.72	3
		Overall		1334.33	1153.33	12.71	3600.00	3.00	1363.67	14.73	53.14	3.00
	60%	5	5	1607	1241	22.78	3600.00	3	1607	22.78	20.01	3
		7	5	1259	1033	17.95	3600.00	3	1263	18.21	1.38	3
		11	5	926	873	5.72	3600.00	3	962	9.25	0.02	3
		Overall		1264.00	1049.00	15.48	3600.00	3.00	1277.33	16.75	7.14	3.00
	70%	5	5	1573	1229	21.87	3600.00	3	1573	21.87	4.60	3
		7	5	1259	1125	10.64	3600.00	3	1341	16.11	2.08	3
		11	5	902	902	0.00	801.30	3	962	6.24	9.99	3
		Overall		1244.67	1085.33	10.84	2667.10	3.00	1292.00	14.74	5.56	3.00
	Overall			1281.00	1095.89	13.01	3289.03	3.00	1311.00	15.40	20.64	3.00

Table 8.4 Results for instances with $25 \leqslant |W| \leqslant 35$

				MILP formulation					BRKGA			
$|W|$	β	c	M_{UB}	UB	LB	gap (%)	Time	M	UB	gap (%)	Time	M
25	50%	5	5	2458	1358	44.75	3600.00	3	2462	44.84	39.10	3
		7	5	2066	1134	45.11	3600.00	3	2067	45.14	2.76	3
		11	5	1715	1042	39.24	3600.00	3	1715	39.24	0.99	3
		Overall		2079.67	1178.00	43.04	3600.00	3.00	2081.33	43.07	14.28	3.00
	60%	5	5	2431	1324	45.54	3600.00	3	2461	46.20	30.93	3
		7	5	2027	1041	48.64	3600.00	3	2027	48.64	41.72	3
		11	5	1604	872	45.64	3600.00	3	1640	46.83	94.10	3
		Overall		2020.67	1079.00	46.61	3600.00	3.00	2042.67	47.22	55.58	3.00
	70%	5	5	2418	1358	43.84	3600.00	3	2368	42.65	2.97	3
		7	5	2044	1094	46.48	3600.00	3	1994	45.14	1.69	3
		11	5	1552	899	42.07	3600.00	3	1577	42.99	4.73	3
		Overall		2004.67	1117.00	44.13	3600.00	3.00	1979.67	43.59	3.13	3.00
	Overall			2035.00	1124.67	44.59	3600.00	3.00	2034.56	44.63	24.33	3.00
30	50%	5	7	–	1610	–	3600.00	–	3183	49.42	38.95	4
		7	7	–	1456	–	3600.00	–	2586	43.70	28.77	4
		11	7	2243	1262	43.74	3600.00	4	2169	41.82	14.73	4
		Overall		2243.00	1442.67	43.74	3600.00	4.00	2646.00	44.98	27.48	4.00
	60%	5	7	–	1593	–	3600.00	–	3149	49.41	93.49	4
		7	7	2585	1525	41.01	3600.00	4	2476	38.41	80.56	4
		11	7	3272	1053	67.82	3600.00	6	2180	51.70	7.56	4
		Overall		2928.50	1390.33	54.41	3600.00	5.00	2601.67	46.51	60.54	4.00
	70%	5	7	–	1559	–	3600.00	–	3102	49.74	161.09	4
		7	7	3375	1229	63.59	3600.00	5	2514	51.11	39.63	4
		11	7	2104	1109	47.29	3600.00	4	1992	44.33	127.46	4
		Overall		2739.50	1299.00	55.44	3600.00	4.50	2536.00	48.39	109.39	4.00
	Overall			2715.80	1377.33	52.69	3600.00	4.60	2594.56	46.63	65.80	4.00
35	50%	5	7	–	1561	–	3600.00	–	4254	63.31	192.99	5
		7	7	–	1454	–	3600.00	–	3553	59.08	11.33	4
		11	7	–	1236	–	3600.00	–	3132	60.54	14.42	4
		Overall		–	1417.00	–	3600.00	–	3646.33	60.97	72.91	4.33
	60%	5	7	–	1655	–	3600.00	–	4009	58.72	47.37	4
		7	7	–	1295	–	3600.00	–	3592	63.95	65.35	4
		11	7	–	1121	–	3600.00	–	2805	60.04	4.27	4
		Overall		–	1357.00	–	3600.00	–	3468.67	60.90	39.00	4.00
	70%	5	7	10,139	1577	84.45	3600.00	6	3651	56.81	234.11	4
		7	7	–	1341	–	3600.00	–	3376	60.28	339.96	5
		11	7	9105	1239	86.39	3600.00	6	2893	57.17	5.02	4
		Overall		9622.00	1385.67	85.42	3600.00	6.00	3306.67	58.09	193.03	4.33
	Overall			9622.00	1386.56	85.42	3600.00	6.00	3473.89	59.99	101.65	4.22

Column "M_{UB}" represents an upper bound of the number of vehicles required to the demand of all customers. The column group "MILP formulation" reports the results obtained by the MILP formulation proposed for ECCTP. In this column group, "time" represents the total execution time of MILP formulation and the symbol "-" means that no upper bounds were obtained. The column group "BRKGA" reports the results obtained by the BRKGA metaheuristic proposed for ECCTP. In this column group, "time" represents the execution time until the best UB was obtained. For all instances, the last row of each variation of β reports the average values of each column. Similarly, the last row of each variation of $|W|$ reports the average values of each column.

The value of parameter M_{UB} can be quite important regarding the optimization difficulty of instances by the proposed methods. If the value is too low, it may be difficult to find any feasible solution. Conversely, a big value can turn the instances unnecessarily difficult to be solved by the MILP formulation since the number of variables relies on M_{UB}. Therefore, we considered a moderate value of M_{UB} for each instance computed by the following formula: $M_{UB} = \lfloor \frac{3}{2} * M_{LB} \rfloor + 1$, where $M_{LB} = \lceil \sum_{i \in W} d_i / Q \rceil$ is a lower bound on the number of vehicles for any feasible solution.

Table 8.3 shows that parameter c is crucial for the difficulty of the instances and cost of optimal solutions. The MILP method addressed more easily instances with $c = 11$, both in solution quality and processing time, when compared to $c \in \{5, 7\}$. For example, the only optimal solution obtained for $|W| = 20$ is the instance with $c = 11$ and $\beta = 70\% * d_G$.

Parameter β also showed a substantial impact on solution costs. The decrease of solution cost when increasing the vehicle batteries capacity from $\beta = 50\% * d_G$ to $\beta = 60\% * d_G$ is substantially greater than from $\beta = 60\% * d_G$ to $\beta = 70\% * d_G$. For example, observing the optimal costs of $|W| = 10$ instances, the average cost decrease from the first increment in battery capacity was approximately 8%, while the second increment incurred in approximately 9% cost reduction. The same pattern was observed for $|W| = 15, 20, 25$. The economic interpretation of these results is that the increase in the battery capacity has diminishing returns concerning transportation costs. This kind of analysis could support the decision-making process for which vehicles model should be used given their battery capacities and potential cost reductions.

Table 8.4 shows the difficulty of optimizing the largest instances $|W| = 30, 35$. For many of these instances, the MILP method did not obtain any feasible solution. This probably is related to the energy constraints since this happened especially more frequently for $\beta = 50\% * d_G$ and less frequently for $\beta = 70\% * d_G$. This suggests that reformulating the problem using a stronger set of constraints to model the battery capacities may be a promising research topic. Conversely, BRKGA was shown to be an effective approach to obtain feasible solutions.

8.7 Conclusion

We proposed a new problem called electric capacitated covering tour problem (ECCTP), which joins two branches in vehicle routing, the green vehicle routing and the covering tour vehicle routing.

The ECCTP was formulated as a mixed integer linear programming (MILP) model. In contrast to other formulations of electric vehicle routing problems, our model does not make any assumptions or restrictions concerning the number of times a vehicle visits the recharging station.

A biased random-key genetic algorithm (BRKGA) metaheuristic was proposed as a methodology to obtain high-quality solutions for large instances. The BRKGA decoder's main features are: (1) a split procedure, which optimally splits a single non-capacitated route into capacitated routes; (2) a dynamic programming procedure that optimally inserts stations visits in each vehicle route.

A set of benchmark instances, adapted from the CVRP literature, was proposed for the ECCTP. Computational experiments showed the effectiveness of the proposed methods. The MILP formulation solved the small instances ($|W| \leqslant 15$) in reasonable processing time (less than 1 h). For bigger instances ($|W| \geqslant 30$), the BRKGA outperformed the MILP, obtaining substantially better solutions while taking less computational time.

The computational experiments also allowed analyzing how the parameters of electric vehicle battery capacity and coverage neighborhood size could impact the optimal solution cost and the difficulty to solve the instances. The proposed methods were found to be sensitive to the neighborhood size concerning the solution gap. Moreover, the analysis emphasized the importance of vehicle battery capacity when addressing the economic feasibility of a transportation plan.

Future studies of the ECCTP should investigate stronger mathematical formulations and also the development of local search methods as well as intensification and diversification strategies.

References

1. Golden, B.: Vehicle routing problems. Technical report. M.I.T. Operations Research Center, Cambridge (1975)
2. Toth, P., Vigo, D.: Vehicle routing: problems, methods, and applications. SIAM, New York (2014)
3. Leggieri, V., Haouari, M.: A practical solution approach for the green vehicle routing problem. Trans. Res. E Logistics Transp. Rev. **104**, 97–112 (2017)
4. Erdoğan, S., Miller-Hooks, E.: A green vehicle routing problem. Transp. Res. E Logistics Transp. Rev. **48**(1), 100–114 (2012). Select Papers from the 19th International Symposium on Transportation and Traffic Theory
5. Allahyari, S., Salari, M., Vigo, D.: A hybrid metaheuristic algorithm for the multi-depot covering tour vehicle routing problem. EJOR **242**(3), 756–768 (2015)
6. Labbé, M., Laporte, G.: Maximizing user convenience and postal service efficiency in post box location. Belgian J. Oper. Res. Stat. Comput. Sci. **26**, 21–35 (1986)

7. Foord, F.: Gambia: evaluation of the mobile health care service in west kiang district (1995)
8. Swaddiwudhipong, W., Chaovakiratipong, C., Nguntra, P., Mahasakpan, P., Lerdlukanavonge, P., Koonchote, S.: Effect of a mobile unit on changes in knowledge and use of cervical cancer screening among rural Thai women. Int. J. Epidemiol. 24(3), 493–498 (1995)
9. Current, J.R., Schilling, D.A.: The covering salesman problem. Transp. Sci. 23(3), 208–213 (1989)
10. Applegate, D.L., Bixby, R.E., Chvatal, V., Cook, W.J.: The Traveling Salesman Problem: A Computational Study (Princeton Series in Applied Mathematics). Princeton University, Princeton (2007)
11. Salari, M., Naji-Azimi, Z.: An integer programming-based local search for the covering salesman problem. Comput. Oper. Res. 39(11), 2594–2602 (2012)
12. Salari, M., Reihaneh, M., Sabbagh, M.S.: Combining ant colony optimization algorithm and dynamic programming technique for solving the covering salesman problem. Comput. Ind. Eng. 83, 244–251 (2015)
13. Gendreau, M., Laporte, G., Semet, F.: The covering tour problem. Oper. Res. 45(4), 568–576 (1997)
14. Dumitrescu, A., Mitchell, J.S.B.: Approximation algorithms for tsp with neighborhoods in the plane. J. Algorithms 48(1), 135–159 (2003)
15. Gulczynski, D.J., Heath, J.W., Price, C.C.: The Close Enough Traveling Salesman Problem: A Discussion of Several Heuristics, pp. 271–283. Springer, Boston (2006)
16. Shuttleworth, R., Golden, B.L., Smith, S., Wasil, E.: Advances in Meter Reading: Heuristic Solution of the Close Enough Traveling Salesman Problem over a Street Network, pp. 487–501. Springer, Boston (2008)
17. Coutinho, W.P., Quirino do Nascimento, R., Pessoa, A.A., Subramanian, A.: A branch-and-bound algorithm for the close-enough traveling salesman problem. INFORMS J. Comput. 28(4), 752–765 (2016)
18. Hachicha, M., Hodgson, M.J., Laporte, G., Semet, F.: Heuristics for the multi-vehicle covering tour problem. Comput. Oper. Res. 27(1), 29–42 (2000)
19. Naji-Azimi, Z., Renaud, J., Ruiz, A., Salari, M.: A covering tour approach to the location of satellite distribution centers to supply humanitarian aid. EJOR 222(3), 596–605 (2012)
20. Flores-Garza, D.A., Salazar-Aguilar, M.A., Ngueveu, S.U., Laporte, G.: The multi-vehicle cumulative covering tour problem. Ann. Oper. Res. 258(2), 761–780 (2017)
21. Koç, Ç., Karaoglan, I.: The green vehicle routing problem: a heuristic based exact solution approach. Appl. Soft Comput. 39, 154–164 (2016)
22. Sherali, H., Adams, W.: A hierarchy of relaxations between the continuous and convex hull representations for zero-one programming problems. SIDMA 3(3), 411–430 (1990)
23. Sherali, H.D., Adams, W.P.: A hierarchy of relaxations and convex hull characterizations for mixed-integer zero—one programming problems. Discrete Appl. Math. 52(1), 83–106 (1994)
24. Montoya, A., Guéret, C., Mendoza, J.E., Villegas, J.G.: A multi-space sampling heuristic for the green vehicle routing problem. Transp. Res. C Emerging Technol. 70, 113–128 (2016)
25. Conrad, R.G., Figliozzi, M.A.: The recharging vehicle routing problem. In: Proceedings of the 2011 Industrial Engineering Research Conference (2011)
26. Wang, Y.-W., Lin, C.-C., Lee, T.-J.: Electric vehicle tour planning. Transp. Res. D Transp. Environ. 63, 121–136 (2018)
27. Bektaş, T., Laporte, G.: The pollution-routing problem. Trans. Res. B Methodol. 45(8), 1232–1250 (2011). Supply chain disruption and risk management
28. Gonçalves, J.F., Resende, M.G.C.: Biased random-key genetic algorithms for combinatorial optimization. J. Heuristics 17(5), 487–525 (2011)
29. Toso, R.F., Resende, M.G.C.: A c++ application programming interface for biased random-key genetic algorithms. Optim. Methods Software 30(1), 81–93 (2015)
30. Prins, C., Labadi, N., Reghioui, M.: Tour splitting algorithms for vehicle routing problems. Int. J. Prod. Res. 47(2), 507–535 (2009)
31. Uchoa, E., Pecin, D., Pessoa, A., Poggi, M., Vidal, T., Subramanian, A.: New benchmark instances for the capacitated vehicle routing problem. EJOR 257(3), 845–858 (2017)

Printed in the United States
by Baker & Taylor Publisher Services